Manuals for the Bench

E. CARAFOLI, G. SEMENZA (Eds.)
Membrane Biochemistry. A Laboratory Manual on Transport
and Bioenergetics (1979), 175 pp, ISBN 3-540-09844-5

A. AZZI, U. BRODBECK, P. ZAHLER (Eds.)
Membrane Proteins. A Laboratory Manual (1981), 256 pp,
ISBN 3-540-10749-5

A. AZZI, U. BRODBECK, P. ZAHLER (Eds.)
Enzymes, Receptors, and Carriers of Biological Membranes.
A Laboratory Manual (1984), 165 pp, ISBN 3-540-13751-3

A. AZZI, L. MASOTTI, A. VECLI (Eds.)
Membrane Proteins. Isolation and Characterization (1986),
181 pp, ISBN 3-540-17014-6

U. BRODBECK, C. BORDIER (Eds.)
Post-translational Modification of Proteins by Lipids.
A Laboratory Manual (1988), 148 pp, ISBN 3-540-50215-7

J. F. T. SPENCER, D. M. SPENCER, J. J. BRUCE
Yeast Genetics. A Manual of Methods (1988)
102 pp, ISBN 3-540-18805-3

N. LATRUFFE, Y. GAUDEMER, P. VIGNAIS, A. AZZI (Eds.)
Dynamics of Membrane Proteins and Cellular Energetics
(1988), 278 pp, ISBN 3-540-50047-2

Dynamics of
Membrane Proteins and
Cellular Energetics

Edited by
Norbert Latruffe, Yves Gaudemer,
Pierre Vignais, Angelo Azzi

With 30 Figures

Springer-Verlag
Berlin Heidelberg New York
London Paris Tokyo

Professor Norbert Latruffe
Université de Franche-Comté
Laboratoire de Biochimie et Biologie Moléculaire
UA CNRS 531
25030 Besançon Cedex – France

Professor Yves Gaudemer
Université de Franche-Comté
Laboratoire de Biochimie et Biologie Moléculaire
UA CNRS 531
25030 Besançon Cedex – France

Professor Pierre Vignais
CENG – DRF Biochimie
Laboratoire de Biochimie
UA CNRS 1130, BP 85 X
38041 Grenoble Cedex – France

Professor Angelo Azzi
Universität Bern
Institut für Biochemie und Molekularbiologie
Bühlstrasse 28
3012 Bern – Switzerland

QP
552
.M44
D96
1988

ISBN 3-540-50047-2 Springer-Verlag Berlin Heidelberg New York
ISBN 0-387-50047-2 Springer-Verlag New York Berlin Heidelberg

Printing and binding: Druckhaus Beltz, Hemsbach/Bergstr.
2131/3130-543210 – Printed on acid-free paper

Preface

This manual on "Dynamics of Membrane Proteins and Cellular Energetics" is the result of a FEBS-CNRS Course held in Grenoble and Besançon in September 1987. It appears to be, after the first, published in 1979 the fifth of the series. After focussing on the "Biochemistry of Membranes" (1979) it was the turn of "Membrane Proteins" (1981) and of Enzymes, Receptors and Carriers of Biological Membranes (1983), followed by "Membrane Proteins : Isolation and Characterization" (1986). Although the central issue has always been the biological membrane, its components and its functions, each manual has put the accent on somewhat different issues corresponding to the most innovative, interesting research in the field.

After almost a decade this new manual appears, which stresses the aspect of the integration of membrane research at a cellular level. Such a novel emphasis is the consequence of the common interest of cell biology and biochemistry to understand the results of the biochemical analysis of membrane proteins in the context of the cell complexity. Consequently most of the experimental protocols are dealing with cellular models, but with clear reference to the function and structure of isolated membrane proteins.

The present manual is divided into two sections. Part one is meant as a theoretical introduction to part two, the experimental section. The theoretical part has been kept very limited not to change the dominating practical character of the manual but it has, in any case, been introduced in order to give the interested scientist not only some simple, well described protocols, but some general background (in form of examples) on the field of membrane proteins and cellular events.

BNA H,5/89

Who should be interested in having this manual in his
bookshelf ? Everyone who likes to set up experiments using
cellular models but lacks previous experience in the field. He
can be the specialist as well, who likes to leave in the hands
of his students a simple, reliable set of protocols, or he can
be the teacher, who likes to run a practical course.

This manual is not encyclopaedic and does not want to compete
with other publications which have this character. Rather than
completeness we have chosen simplicity, rather than giving
principles we have chosen to give examples.

A. Azzi
(for the editors)

Contents

Contributors

ADAMI Pascale, Université de Franche-Comté, Laboratoire de Biochimie et Biologie Moléculaire, UA CNRS 531, 25030 BESANCON Cedex - FRANCE

AZZI Angelo, University of Bern, Institute of Biochemistry and Molecular Biology, Bühlstrasse 28, CH 3012 BERN - SWITZERLAND

BAGGIOLINI Marco, Theodor Kocher Institute, University of Bern, Freiestrasse 1, CH 3000 BERN 9 - SWITZERLAND

BAILLY Anne, Université de Franche-Comté, Laboratoire de Biochimie et Biologie Moléculaire, UA CNRS 531, 25030 BESANCON Cedex - FRANCE

BERREZ Jean-Marc, Université de Franche-Comté, Laboratoire de Biochimie et Biologie Moléculaire, UA CNRS 531, 25030 BESANCON Cedex - FRANCE

BOF Mireille, CENG Laboratoire de Biochimie, DRF/Biochimie, BP 85 X, 38041 GRENOBLE Cedex - FRANCE

CHERKAOUI MALKI Mustapha, Université de Franche-Comté, Laboratoire de Biochimie et Biologie Moléculaire, UA CNRS 531, 25030 BESANCON Cedex - FRANCE

DELAAGE Michel, Immunotech Luminy, Case 915, 13288 MARSEILLE Cedex 9 - FRANCE

DERANLEAU D.A., Theodor Kocher Institute, University of Bern, Freiestrasse 1, CH 3000 BERN 9 - SWITZERLAND

DE VEYRAC Béatrice, Immunotech Luminy, Case 915, 13288 MARSEILLE Cedex 9 - FRANCE

DEWALD Béatrice, Theodor Kocher Institute, University of Bern, Freiestrasse 1, CH 3000 BERN 9 - SWITZERLAND

DOUCE Roland, CENG, Laboratoire de Physiologie Cellulaire Végétale, BP 85 X, 38041 GRENOBLE Cedex - FRANCE

DOUSSIERE Jacques, CENG Laboratoire de Biochimie, DRF/Biochimie, BP 85 X, 38041 GRENOBLE Cedex - FRANCE

GONZALEZ C., CENG Laboratoire de Biochimie, DRF/Biochimie, BP
    85 X, 38041 GRENOBLE Cedex - FRANCE
JOB Didier, CENG INSERM U 244, LBio/DRF, BP 85 X, 38041
    GRENOBLE Cedex - FRANCE
JOYARD Jacques, CENG, Laboratoire de Physiologie Cellulaire
    Végétale, BP 85 X, 38041 GRENOBLE Cedex - FRANCE
KANTE Arlette, Université de Franche-Comté, Laboratoire de
    Biochimie et Biologie Moléculaire, UA CNRS 531, 25030
    BESANCON Cedex - FRANCE
KLEIN Gérard, CENG Laboratoire de Biochimie, DRF/Biochimie, BP
    85 X, 38041 GRENOBLE Cedex - FRANCE
KURKDJIAN Armen, Laboratoire de Physiologie Végétale, CNRS -
    Avenue de la terrasse, BP 1, 91190 GIF/YVETTE - FRANCE
LATRUFFE Norbert, Université de Franche-Comté, Laboratoire de
    Biochimie et Biologie Moléculaire, UA CNRS 531, 25030
    BESANCON Cedex - FRANCE
LE QUÔC Danielle, Université de Franche-Comté, Laboratoire de
    Biochimie et Biologie Moléculaire, UA CNRS 531, 25030
    BESANCON Cedex - FRANCE
LE QUÔC Khanh, Université de Franche-Comté, Laboratoire de
    Biochimie et Biologie Moléculaire, UA CNRS 531, 25030
    BESANCON Cedex - FRANCE
LUTHY Roland, University of Bern, Institute of Biochemistry
    and Molecular Biology, Bühlstrasse 28, CH 3012 BERN -
    SWITZERLAND
MAHONEY Charles, University of Bern, Institute of Biochemistry
    and Molecular Biology, Bühlstrasse 28, CH 3012 BERN -
    SWITZERLAND
MIDOUX Patrick, Centre de Biophysique Moléculaire du CNRS, 1,
    rue Haute, 45071 ORLEANS Cedex - FRANCE
MILANI Daria, University of Padova, Institute di Patologia
    Generale, Via Loredan, 35131 PADOVA - ITALY
MONSIGNY Michel, Centre de Biophysique Moléculaire du CNRS, 1,
    rue Haute, 45071 ORLEANS Cedex - FRANCE
MOREL Anne, Immunotech Luminy, Case 915, 13288 MARSEILLE Cedex
    9 - FRANCE

MOREL Françoise, CENG Laboratoire de Biochimie, DRF/Biochimie
    BP 85 X, 38041 GRENOBLE Cedex - FRANCE

NEUBURGER Michel, CENG, Laboratoire de Physiologie Cellulaire
    Végétale, BP 85 X, 38041 GRENOBLE Cedex - FRANCE

PIROLLET Fabienne, CENG INSERM U 244, LBio/DRF, BP 85 X, 38041
    GRENOBLE Cedex - FRANCE

POZZAN Tulio, University of Ferrara, Institute di Patologia
    Generale, FERRARA - ITALY

PUGIN Alain, Université de Franche-Comté, Laboratoire de
    Biochimie , UA CNRS 531, 25030 BESANCON Cedex - FRANCE

ROLS Marie-Pierre, Centre de Recherches en Biochimie et
    Génétique Cellulaires du CNRS, 118, route de Narbonne,
    31062 TOULOUSE Cedex - FRANCE

SATRE Michel, CENG Laboratoire de Biochimie, DRF/Biochimie, BP
    85 X, 38041 GRENOBLE Cedex - FRANCE

SEMENZA Giorgio, ETH-Zentrum, Laboratorium für Biochemie,
    CH-8092 ZURICH - SWITZERLAND

TEISSIE Justin, Centre de Recherches en Biochimie et
    Génétique Cellulaires du CNRS, 118, route de Narbonne,
    31062 TOULOUSE Cedex - FRANCE

THELEN M., Theodor Kocher Institute, University of Bern,
    Freiestrasse 1, CH 3000 BERN 9 - SWITZERLAND

Von TSCHARNER V., Theodor Kocher Institute, University of
    Bern,
    Freiestrasse 1, CH 3000 BERN 9 - SWITZERLAND

TREVES Susan, University of Padova, Institute di Patologia
    Generale, Via Loredan, 35131 PADOVA - ITALY

VIGNAIS Pierre, CENG Laboratoire de Biochimie, DRF/Biochimie,
    BP 85 X, 38041 GRENOBLE Cedex - FRANCE

WYMANN M.P., Theodor Kocher Institute, University of Bern,
    Freiestrasse 1, CH 3000 BERN 9 - SWITZERLAND

PART ONE : THEORETICAL INTRODUCTION

THE STRUCTURE AND FUNCTION OF Ca^{+2} - AND PHOSPHOLIPID-DEPENDENT PROTEIN KINASE (PROTEIN KINASE C), A TRANS-MEMBRANE SIGNAL TRANSDUCER

C.W. MAHONEY & A. AZZI

Abbreviations used : PKC, protein kinase c ; PS, phosphatidyl-serine ; PDB, phorbol 12, 13 dibutyrate ; DAG, diacylglyce-rol ; M_r, apparent molecular weight ; kDa, kilo-dalton.

Ca^{+2}- and phospholipid- dependent protein kinase (protein kinase c), first discovered by Nishizuka's group as a protease-activated kinase in 1977 (Takai et al., 1977 ; Inoue et al., 1977), has generated tremendous interest in the biochemical community in the last 11 years because of its implication in numerous biological processes including tumor promotion (Hecker & Schmidt, 1979), membrane transporter and channel modulation (Sigel et al., 1988 ; Costa & Catterall, 1985 ; Liles et al., 1986 ; Witters et al., 1985), differen-tiation (Morin et al., 1987 ; Pahlman et al., 1983), muscle contraction (Ikeba et al., 1985 ; Nishikawa et al., 1984 & 1985), neural synaptic communication (Tanaka et al., 1986), secretion (Ieyasu et al., 1982 ; Kajikawa et al., 1983), the respiratory burst (Mahoney & Azzi, in press, 1988 ; Lüthy & Azzi, 1987 ; Mahoney et al., 1986 ; Serhan et al., 1983 ; Fujita et al., 1984), the immune response (Patel et al., 1987), growth (Cooper et al., 1982 ; Cochet et al., 1984 ; McCaffrey et al., 1984 ; Davis & Czech, 1985) and platelet aggregation (Froscio et al., 1988 ; Mahoney & Azzi, in press ; Naka et al., 1983 ; Sano et al., 1983 ; Watson et al., 1988 ; Mahoney et al., this volume) (for reviews see Nishizuka 1984 & 1986). This paper will focus on the basic structural features of protein kinase c and some of the most recent developments in the elucidation of its structure and function.

Protein kinase c (PKC) consists of a single polypep-tide with an apparent molecular weight (M_r) of 84,000 kDa and

is the only protein kinase examined to date that requires Ca^{+2}
and phospholipid for activation, in contrast to the other well
characterized cAMP-activated (for review see Beavo & Mumby,
1982), cGMP-activated (for review see Kua & Shoji, 1982) and
the Ca^{+2} - calmodulin-activated protein kinases (for review
see Schulman, 1982). An additional feature that distinguishes
protein kinase c from the other protein kinases is its
activation by tumor promoters (e.g. the phorbol 12, 13
diesters) and the neutral lipid diacylglycerol (DAG), both of
which can lower the Ca^{+2} requirement for enzyme activation to
the sub-micromolar level (Takai et al., 1979 ; Castagna et
al., 1982). PKC is a cytosolic protein in the cell resting
state ($\simeq$ 80 % cytosolic, $\simeq$ 20 % membrane associated) (Ashendel
et al., 1983) and upon cell stimulation by a variety of
extracellular ligands (e.g. hormones, neurotransmitters, the
phorbol 12, 13 diesters or DAG) most of the PKC becomes cell
membrane associated ($\simeq$ 80 % membrane associated, $\simeq$ 20 %
cytosolic) (Kraft & Anderson, 1983 ; Kraft et al., 1982). A
rise in internal Ca^{+2} alone (Melloni et al., 1985 ; Wolf et
al., 1985 b) can cause this redistribution and activation of
PKC as can phorbol diester or DAG (Gopalakrishna et al.,
1986 ; Wolf et al., 1985 b). Physiologically, in many cases,
when an extracellular ligand (as above) binds to its cell
surface receptor, a phospholipase C is activated through a
coupling with G protein(s) thereby resulting in the hydrolysis
of phosphatidylinositoldiphosphate (PIP2) to inositoltriphos-
phate (IP3) and DAG. The water soluble IP3 can indirectly
activate PKC by inducing a release of intracellularly stored
Ca^{+2} (non-mitochondrial) (Berridge, 1984 ; Michell, 1983) and
DAG can directly induce a redistribution and activation of the
enzyme (Gopalakrishna et al., 1986 ; Wolf et al., 1985 b).
Binding of radioactive phorbol 12, 13 diester can be competitively inhibited by the addition of DAG suggesting that both
compounds have overlapping binding sites on the enzyme-phos-
pholipid complex (Sharkey & Blumberg, 1985 ; Sharkey et al.,
1984).

The Catalytic and Regulatory Domains

PKC is readily cleaved by a Ca^{+2} activated intra-cellular protease into a 50 kDa catalytically active (Ca^{+2} - phospholipid - independent) and a 30 kDa M_r regulatory fragment (Inoue et al., 1977 ; Melloni et al., 1985 ; Nakadate et al., 1987). There is some evidence to suggest that proteolysis is the mechanism of down regulation of protein kinase c (Young et al., 1987 ; Woodgett & Hunter, 1987). The 50 and 30 kDa M_r catalytic and regulatory fragments can also be generated in vitro by limited tryptic digestion (Lee & Bell, 1986 ; Mochly-Rosen & Koshland, 1987 ; Huang & Huang, 1986) which has been useful in elucidating the function of these domains. Recently, Lee & Bell (1986) have presented evidence demonstrating that the Ca^{+2}, phosphatidylserine (PS), and PDB (Huang & Huang, 1986) binding sites are located on the regulatory fragment. The 30 and the 50 kDa regulatory and catalytic domains have been localized to the N and C terminal regions of the protein respectively (Parker et al., 1986). Since most of the regulators of PKC activity are hydrophobic compounds which compete for PDB or PS binding it is likely that these compounds also modulate the activity of PKC by binding in the 30 kDA M_r domain. Current knowledge suggests that the active site of resting PKC located in the 50 kDa domain is blocked by the 30 kDa regulatory domain and that on the binding of Ca^{+2}, PS, and PDB or DAG in this domain the active site becomes reactive to the substrates MgATP and protein acceptor.

Primary Structure, Properties, and Distribution of the α, β, and γ Forms of PKC

Recently the complete nucleic acid sequences for the 3 isozymic forms of PKC (α, β, γ) and their deduced amino acid sequences have been reported (Parker et al., 1986 ; Coussens et al., 1986) thereby facilitating further elucidation of the mechanisms of activation and inhibition by various ligands. The primary sequences of the α, β, γ forms of PKC consist of 672, 673, and 693 amino acid residues and each contains an amino terminus cysteine-rich domain, a down stream predicted Ca^{+2} binding domain, and a conserved protein kinase active site domain in the carboxy terminus (Parker et al., 1986). The differential role of the three isozymes remains unclear yet recent studies have reported differences in the enzymatic properties and distributions of these isozymes. The α, β, γ forms of PKC, readily separated by hydroxylapatite chromatography (Ido et al., 1987 ; Huang, K. et al., 1986 b ; Huang F. et al., 1987 ; Shearman et al., 1987) are differentially activated by unsaturated free fatty acids (Sekiguchi et al., 1987). The γ isoform shows a slight activation (30-40 % relative to Ca^{+2}, PS, DAG activated) in the presence of arachidonic, oleic, or linoleic acid in the 25-100 μM concentration range, whereas the α isoform is activated by these 3 fatty acids (50-400 μM range) in the presence of Ca^{+2} to a similar degree as in the presence of Ca^{+2}, PS, and DAG. The β isoform shows intermediate activation properties by free fatty acid in the presence of Ca^{+2} (Sekiguchi et al., 1987). Although Sekiguchi et al. were able to find no significant difference in the Ca^{+2}, PS, DAG activation properties of the 3 isozymes, in contrast Huang K. et al. (1986 b) found similar Ca^{+2}, PS, DAG activation properties (> 15 fold) for α and β PKC, but only a 4-8 fold activation for the γ form. Ido et al. (1987) have provided preliminary evidence that the α PKC is able to phosphorylate the EGF receptor in membranes of epidermal carcinoma cells the most rapidly, β PKC at an intermediate rate, and γ with the lowest rate. Auto-phosphorylation of α and γ PKC occurs only at serine residues yet in the case of β PKC threonine residue(s) can be phosphorylated as well (Huang K. et al., 1986 b). Since only 1-2 moles Pi/mole PKC are autoincorporated this suggests that in the case of α and γ PKC two serine residues are autophosphorylated whereas for β PKC a single serine and threonine are modified. The differential distribution of α, β, γ PKC isoforms within the brain has been reported by several groups (Huang F. et al., 1987 ; Shearman et al., 1987). Human neutrophils (Sekiguchi et al., 1987) and normal and ras-transformed 3T3 fibroblasts (McCaffrey et al., 1987) contain only α PKC. McCaffrey et al. (1987), in addition, were not able to find any differences between α PKC from normal and ras-transformed 3T3 fibroblasts.

Auto-phosphorylation

Protein kinase c can be auto-phosphorylated (Lepeuch et al., 1983 ; Huang K. et al., 1986 a ; Mochly-Rosen & Koshland, 1987) in both the regulatory and catalytic domains (Newton & Koshland, 1987 ; Huang K. et al., 1986 a) and enzymatic activity is stimulated as a result (Huang K. et al., 1986 a ; Mochly-Rosen & Koshland, 1987). Auto-phosphorylation resulted in an activation of PKC through a 2 fold decrease in the Km for histone (Mochly-Rosen & Koshland, 1987), a 4 fold lowering of the Ca^{+2} requirement for activation (7.5 vs. 31.6 μM), and a 2 fold lowering of the Kd for PDB (Huang K. et al., 1986 a). In addition it was found that the limited-tryptic generated 50 kDa fragment is not capable of auto-phosphorylation and that auto-phosphorylation is an intra-molecular process (Newton & Koshland, 1987 ; Mochly-Rosen & Koshland, 1987). Hence, it appears that auto-phosphorylation activates PKC, yet a report by Wolf et al. (1985 a) provides suggestive evidence that down regulation of PKC may occur through auto-phosphorylation by stimulating dissociation of the active membrane bound enzyme from the membrane. It is quite possible that both mechanisms are working in a temporal sequence of events in vivo.

Inhibitors and Activators of PKC

Several other activators of PKC, namely teleocidin, aplysiatoxin, mezerein, bryostatins, and CC14, in addition to Ca^{+2}, PS, DAG and PDB, have been reported. In all cases, with the exception of CC14, the activator competes with PDB or DAG (Jeffrey & Liskamp, 1986 ; Berkow & Kraft, 1985). In the case of CC14, a well known tumor inducing agent, a 2-3 fold activation was found at several hundreds of micromolar levels of CC14 yet no competition was observed with respect to PDB (Rhogani et al., 1987). Photoaffinity analogs of phorbol, phorbol 12 azidobenzoate, 13-benzoate (Delclos et al., 1983) and DAG, azido-nitro-phenyl- DAG (Lüthy & Azzi, 1987) have been synthesized and utilized to probe the PDB and DAG interaction sites with PKC. In both cases, after photo-incorporation, all of the analog was associated with phospholipid suggesting a complex interaction of Ca^{+2}, PS, and PDB/DAG with the enzyme. It is known that Ca^{+2} and PS are required for phorbol 12, 13 diester binding to holo-PKC (Ashendel et al., 1983). Numerous compounds have been found to inhibit PKC (Kuo et al., 1984 ; Mahoney & Azzi, in press & 1988 ; Mahoney et al., in press) most of which compete at the hydrophobic binding site for PDB/DAG. Recently staurosporine and its secondary amine derivatives have been shown to be the most potent inhibitors (IC_{50} = 3-167 nM) of PKC (Tamaoki et al., 1986 ; Mahoney et al., in press) yet they do not compete for any of the other ligands of PKC (Ca^{+2}, PS, PDB/DAG, histone, MgATP). In addition these compounds inhibit PDB-stimulated platelet aggregation (Mahoney et al., this volume ; Watson et al., 1988) and PDB-specific platelet protein (40 and 20 kDa M_r) phosphorylation (Watson et al., 1988). The fluorescent properties of the staurosporines (Mahoney et al., in press) should make them useful probes in structural as well as cell studies. Other recent inhibitors that have been described include sphingosine and lysosphingolipids (Hannun & Bell, 1987), potential physiological-pathological regulators, which compete with PDB/DAG . The Au(I) anti-rheumatic drugs, Auranofin, Au-S-glucose, and Au-S-malate, inhibit the kinase activity of purified PKC (Mahoney & Azzi, in press & 1988) (IC_{50} = 3-361 µM). In contrast to this, Froscio et al. (1988) have reported that Auranofin stimulates phosphorylation of proteins (40 and 20 kDa M_r) that are specifically phosphory-lated in the presence of phorbol diester (TPA) in intact platelets. It may be that the local milieu of PKC in the intact platelet is sufficiently different (e.g. Auranofin does not induce cytosolic to membrane bound translocation of PKC in contrast to the phorbol diesters, Froscio et al., 1988) to allow for activation by Auranofin rather than inhibition as is the case with purified PKC. Alternatively, in the case of intact platelets, Auranofin may stimulate phosphorylation through inhibition of a phosphatase, through another kinase, or other intermediary metabolic step(s). In addition to small organic inhibitors, some polypeptides and proteins are able to inhibit PKC. Hucho et al. (1987) have reported the isolation of a 40 kDa M_r protein from bovine brain cytosol which inhibits PKC kinase activity. Earlier reports indicate that a 17 kDa Ca^{+2} binding protein from bovine brain (McDonald & Walsh, 1985) and calmodulin, S100, and troponin c can inhibit PKC activity (Albert et al., 1984). House & Kemp (1987) have elegantly demonstrated that a synthetic peptide corresponding to residues 19-36 in PKC (part of the regulatory domain), which resembles a substrate phosphorylation site in its distribution of basic residues, can potently and specifically inhibit (Ki = 147 nM) PKC exogenous and auto-phosphorylation reactions. In addition, on substituting alanine 25 with a serine residue, the peptide becomes a phosphate acceptor peptide. Physiologically residues 19-36 within PKC may play an inhibitory - regulatory role.

References

Albert K.A., Wu W., Naim A.C. & Greengard P. (1984). Inhibition of calmodulin of calcium/phospholipid dependent protein kinase phosphorylation. **Proc. Natl. Acad. Sci. (USA) 81** : 3622-3625.

Ashendel C.L., Staller J.M. & Boutwell R.K. (1983). Identification of a calcium- and phospholipid-dependent phorbol ester binding activity in the soluble fraction of mouse tissues. **Biochem. Biophys. Res. Commun. 111** : 340-345.

Beavo J.A. & Mumby M.C. (1982). Cyclic AMP-dependent protein phosphorylation, in Cyclic Nucleotides, Part 1, Biochemistry. Nathanson J.A. & Kebabian J.W. (eds). Springer Verlag, Berlin. pp. 363-392.

Berkow R.L. & Kraft A.S. (1985). Bryostatin, a non-phorbol macrocyclic lactone, activates intact human polymorphonuclear leukocytes and binds to the phorbol ester receptor. **Biochem. Biophys. Res. Commun. 131** : 1109-1116.

Berridge M.J. (1984). Inositol triphosphate and diacylglycerol as second messengers. **Biochem. J. 220** : 345-360.

Brandt S.J., Niedel J.E., Bell R.M. & Young W.S. (1987). Distinct patterns of expression of different protein kinase c mRNAs in rat tissues. **Cell 49** : 57-63.

Castagna M., Takai Y., Kaibuchi K., Sano K., Kikkawa U. & Nishizuka Y. (1982). Direct activation of calcium-activated, phospholipid-dependent protein kinase by tumor-promoting phorbol esters. **J. Biol. Chem. 257** : 7847-7851.

Cochet C., Gill G.N., Meisenhelder J., Cooper J.A. & Hunter T. (1984). C-kinase phosphorylates the epidermal growth factor receptor and reduces its epidermal growth factor-stimulated tyrosine protein kinase activity. **J. Biol. Chem. 259** : 2553-2558.

Costa M.R. & Catterall W.A. (1985). Phosphorylation of the alpha subunit of the sodium channel by protein kinase c. **Cell. Mol. Neurobiol. 4** : 291-297.

Cooper R.A., Brunwald A.D. & Kuo A.L. (1982). Phorbol ester induction of leukemic cell differentiation is a membrane-mediated process. **Proc. Natl. Acad. Sci.** (USA) **79** : 2865-2869.

Coussens L., Parker P.J., Rhee L., Yang-Feng T.L., Chen E., Waterfield M.D., Francke U. & Ullrich A. (1986). Multiple, distinct forms of bovine and human protein kinase c suggest diversity in cellular signaling pathways. **Science 233** : 859-866.

Davis R.J., & Czech M.P. (1985). Platelet-derived growth factor mimics phorbol diester action on epidermal growth factor receptor phosphorylation at thr 654. **Proc. Natl. Acad. Sci.** (USA) **82** : 4080-4084.

Delclos K.B., Yeh E. & Blumberg P.M. (1983). Specific labeling of mouse brain membrane phospholipids with [20- ^{3}H] phorbol 12-p-azidobenzoate 13-benzoate, a photolabile phorbol ester. **Proc. Natl. Acad. Sci.** (USA) **80** : 3054-3058.

Froscio M., Solanki U., Murray A.N. & Hurst N.P. (1988). Auranofin enhances phosphorylation of purative substrates of protein kinase c in human platelets. **Biochem. Pharmacol. 37** : 366-368.

Fujita I., Irita K., Takeshiga K., & Minakami S. (1984). Diacylglycerol, 1-oleoyl-2-acetylglycerol, stimulates superoxide generation from human neutrophils. **Biochem. Biophys. Res. Commun. 120** : 318-324.

Gopalakrishna R., Barsky S.H., Thomas T.P. & Anderson W.B. (1986). Factors influencing chelator-stable, detergent-extractable, phorbol diester-induced membrane association of protein kinase c. **J. Biol. Chem. 261** : 16438-16445.

Hansson A., Serhan C.N., Haeggstrom J., Ingelman-Sundberg M. & Samuelsson B. (1986). Activation of protein kinase c by lipoxin A and other eicosanoids. Intracellular action of oxygenation products of arachidonic acid. **Biochem. Biophys. Res. Commun. 134** : 1215-1222.

Hecker E. & Schmidt R. (1979). Phorbolesters - the irritants and cocarcinogens of Croton tiglium L. Forstchr. Chem. Org. Naturstoff 31 : 377-467.

House C. & Kemp B.E. (1987). Protein kinase c contains a pseudosubstrate prototype in its regulatory domain. Science 238 : 1726-1728.

Huang F.L., Yoshida Y., Nakabayashi H. & Huang K. (1987). Differential distribution of protein kinase c isozymes in the various regions of brain. J. Biol. Chem. 262 : 15714-15720.

Huang K., Chan K.J., Singh T.O., Nakabayashi H. & Huang F.L. (1986 a). Autophosphorylation of rat brain Ca^{+2} -activated and phospholipid -dependent protein kinase. J. Biol. Chem. 261 : 12134-12140.

Huang K. & Huang F.L. (1986). Conversion of protein kinase c from a Ca^{+2} -dependent to an independent form of phorbol ester -binding protein by digestion with trypsin. Biochem. Biophys. Res. Commun. 139 : 320-326.

Huang K., Nakabashi H. & Huang F.L. (1986 b). Isozymic forms of rat brain Ca^{+2} -activated and phospholipid -dependent protein kinase. Proc. Natl. Acad. Sci. (USA) 83 : 8535-8539.

Hucho F., Kruger H., Pribilla I. & Oberdieck U. (1987). A 40 kDa inhibitor of protein kinase c purified from bovine brain. FEBS Lett 211 : 207-210.

Ido M., Sekiguchi K., Kikkawa U. & Nishizuka Y. (1987). Phosphorylation of the epidermal growth factor receptor from A431 epidermoid carcinoma cells by three distinct types of protein kinase c. FEBS Lett 219 : 215-218.

Iekeba M., Inagaki M., Kanamaru K. & Hidaka H. (1985). Phosphorylation of smooth muscle myosin light chain kinase by Ca^{+2} -activated, phospholipid -dependent protein kinase. J. Biol. Chem. 260 : 4547-4550.

Ieyasu H., Takai Y., Kaibuchi K., Sawamura M. & Nishizuka Y. (1982). A role of calcium - activated, phospholipid -dependent protein kinase in platelet-activating factor-induced serotonin release from rabbit platelets. **Biochem. Biophys. Res. Commun. 108** : 1701-1708.

Inoue M., Kishimoto A., Takai Y. & Nishizuka Y. (1977). Studies on a cyclic nucleotide - independent protein kinase and its proenzyme in mammalian tissues. **J. Biol. Chem. 252** : 7610-7616.

Jeffrey A.M. & Liskamp R.M.J. (1986). Computer -assisted molecular modeling of tumor promoters : rationale for the activity of phorbol esters, teleocidin B, and aplysiatoxin. **Proc. Natl. Acad. Sci.** (USA) **83** : 241-245.

Kajikawa S., Kaibuchi K., Matsubara T., Kikkawa U., Takai Y. & Nishizuka Y. (1983). A possible role of protein kinase c in signal-induced lysosomal enzyme release. **Biochem. Biophys. Res. Commun. 116** : 743-760.

Kraft A.S. & Anderson W.B. (1983). Phorbol esters increase the amount of Ca^{+2}, phospholipid-dependent protein kinase associated with plasma membrane. **Nature 301** : 621-623.

Kraft A.S., Anderson W.B., Cooper H.L. & Sando J.J. (1982). Decrease in cytosolic calcium/phospholipid -dependent protein kinase activity following phorbol ester treatment of EL4 thyoma cells. **J. Biol. Chem. 257** : 13193-13196.

Kuo J.F., Schatzman R.C., Turner R.S. & Mazzei G.J. (1984). Phospholipid -sensitive Ca^{+2} -dependent protein kinase : a major protein phosphorylation system. **Mol. & Cell. Endocrinol. 35** : 65-73.

Kuo J.F. & Shoji M. (1982). Cyclic GMP -dependent protein phosphorylation, in Cyclic Nucleotides, Part 1, Biochemistry. Nathanson J.A. & Kababian J.W. (eds). Springer Verlag, Berlin. pp. 393-424.

Lee M. & Bell R.M. (1986). The lipid binding, regulatory domain of protein kinase c. **J. Biol. Chem. 261** : 14867-14870.

Lepeuch C.J., Ballester R. & Rosen O.M. (1983). Purified rat brain calcium- and phospholipid-dependent protein kinase phosphorylates ribosomal protein S6. **Proc. Natl. Acad. Sci.** (USA) **80** : 6858-6862.

Liles W.C., Hunter D.D., Meier K.E. & Nathanson N.M. (1986). Activation of protein kinase c induces rapid internalization and subsequent degradation of muscarinic acetylcholine receptors in neuroblastoma cells. **J. Biol. Chem. 261** : 5307-5313.

Lüthy R. & Azzi A. (1987). Fluorescent and photoactive probes for the study of protein kinase c. **Eur. J. Biochem. 162** : 387-391.

Mahoney C.W. & Azzi A. (1988). The Au(I) anti-rheumatic compounds inhibit protein kinase c. **Experientia 44**, A82.

Mahoney C.W. & Azzi A. (in press). Ca^{+2} - and phospholipid dependent protein kinase (protein kinase c) : a pleiotropic signal protein and its inhibition by staurosporine and auranofin, in **Perspectives in Molecular Approaches to Human Diseases**. Papa S. (ed.). Ellin Horwood, London.

Mahoney C.W., Fredenhagen A., Peter H. & Azzi A. (in press). Staurosporine secondary amine derivatives : synthesis, fluorescent properties, and inhibitory action protein kinase c. **Biological Chem.** Hoppe-Seyler.

Mahoney C.W., Lüthy R. & Azzi A. (1986). Membrane signal transduction via protein kinase c, in Membrane Proteins. Azzi A. et al. (eds.) Springer Verlag, Berlin. pp. 56-66.

Malkinson A.M., Girard P.R. & Kuo J.F. (1987). Strain -specific postnatal changes in the activity and tissue levels of protein kinase c. **Biochem. Biophys. Res. Commun. 145** : 733-739.

McCaffrey P.G., Friedman B.A. & Rosner M.R. (1984). Diacylglycerol modulates binding and phosphorylation of the epidermal growth factor receptor. **J. Biol. Chem. 259** : 12502-12507.

McCaffrey P.G., Rosner M.R., Kikkawa U., Sekiguchi K., Ogita K., Ase K. & Nishizuka Y. (1987). Characterization of protein kinase c from normal and transformed cultured murine fibroblasts. **Biochem. Biophys. Res. Commun.** **146** : 140-146.

Mc Donald J.R, & Walsh M.P. (1985). Ca^{+2}-binding proteins from bovine brain including a potent inhibitor of protein kinase c. **Biochem J.** **232** : 559-567.

Melloni E., Pontremolli S., Michetti M., Sacco O., Sparatore B., Salamino F. & Horrecker B.L. (1985). Binding of protein kinase c to neutrophil membranes in the presence of Ca^{+2} and its activation by a Ca^{+2} -requiring proteinase. **Proc. Natl. Acad. Sci.** (USA) **82** : 6435-6439.

Michell R. (1983). Ca^{+2} and protein kinase c : two synergistic cellular signals. **Trends Biochem. Sci.** **8** : 263-265.

Mochly-Rosen D. & Koshland D.E. (1987). Domain structure and phosphorylation of protein kinase c. **J. Biol. Chem.** **262** : 2291-2297.

Mori T., Takai Y., Yu B., Takahashi J., Nishizuka Y. & Fujikura T. (1982). Specificity of the fatty acyl moieties of diacylglycerol for the activation of calcium -activated phospholipid-dependent protein kinase. **J. Biochem.** **91** : 427-431.

Morin M.J., Kreutter D., Rasmussen H. & Sartorelli A.C. (1987). Disparate effects of activators of protein kinase c on HL-60 promyelocytic leukemia cell differentiation. **J. Biol. Chem.** **262** : 11758-11763.

Murakami K. & Routtenberg A. (1985). Direct activation of purified protein kinase c by unsaturated fatty acids (oleate and arachidonate) in the absence of phospholipids and Ca^{+2}. **FEBS Lett.** **192** : 189-193.

Naka M., Nishikawa M., Adelstein R.S. & Hidaka H. (1983).
Phorbol ester -induced activation of human platelets is
associated with protein kinase c phosphorylation of myosin
light chains. **Nature** 306 : 490-492.

Nakadate T., Jeng A.Y. & Blumberg P.M. (1987). Effect of
phospholipid on substrate phosphorylation by a catalytic
fragment of protein kinase c. **J. Biol. Chem.** 262 : 11507-
11513.

Newton A.C. & Koshland D.E. (1987). Protein kinase c
autophosphorylates by an intrapeptide reaction. **J. Biol.
Chem.** 262 : 10185-10188.

Nishikawa M., Shirakawa S. & Adelstein R.S. (1985).
Phosphorylation of smooth muscle myosin light chain kinase
by protein kinase c : comparative study of the phosphorylated sites. **J. Biol. Chem.** 260 : 8978-8983.

Nishikawa M., Seller J.R., Adelstein R.S. & Hidaka H. (1984).
Protein kinase c modulates in vitro phosphorylation of the
smooth muscle heavy meromyosin by myosin light chain kinase
J. Biol. Chem. 259 : 8808-8814.

Nishizuka Y. (1984). The role of protein kinase c in cell
surface signal transduction and tumor promotion. **Nature**
308 : 693-698.

Nishizuka Y. (1986). Studies and perspectives of protein
kinase c. **Science** 233 : 305-311.

Pahlman S., Ruusala A., Abrahamsson L., Odelstad L. & Nillson
K. (1983). Kinetics and concentration effects of TPA
-induced differentiation of cultured neuroblastoma cells.
Cell Differentiation 12 : 165-170.

Parker P.J., Coussens L., Totty N., Rhee L., Young S., Chen
E., Stabel S., Waterfield M.D. & Ullrich A. (1986). The
complete primary structure of protein kinase c - the major
phorbol ester receptor. **Science** 233 : 853-858.

Patel M.D., Samelson L.E. & Klausner R.D. (1987). Multiple kinases and signal transduction : phosphorylation of the T cell antigen receptor complex. **J. Biol. Chem.** **262** : 5831-5838.

Rhogani M., DaSilva C. & Castagna M. (1987). Tumor promoter chloroform is a potent protein kinase c activator. **Biochem. Biophys. Res. Commun.** **142** : 738-744.

Safran A., Neumann D. & Fuchs S. (1986). Analysis of acetylcholine receptor phosphorylation sites using antibodies to synthetic peptides and monoclonal antibodies. **EMBO J. 5** : 3175-3178.

Sano K., Takai Y., Yamanashi J. & Nishizuka Y. (1983). A role of calcium -activated phospholipid-dependent protein kinase in human platelet activation. **J. Biol. Chem.** **258** : 2010-2013.

Schulman H. (1982). Calcium -dependent protein phosphorylation, in Cyclic Nucleotides, Part 1, Biochemistry. Nathanson J.A. & Kebabian, J.W. (eds.) Springer Verlag, Berlin. pp. 425-478.

Sekiguchi K., Tsukuda M., Ogita K., Kikkawa U. & Nishizuka Y. (1987). Three distinct forms of rat brain protein kinase c : differential response to unsaturated fatty acids. **Biochem. Biophys. Res. Commun. 145** : 797-802.

Serhan C.N., Broekman M.D., Korchak H.M., Smolen J.E., Marcus A.J. et al. (1983). Changes in phosphatidylinositol and phosphatidic acid in stimulated human neutrophils. **Biochim. Biophys. Acta 762** : 420-428.

Sharkey N.A. & Blumberg P.M. (1985). Kinetic evidence that 1,2-diolein inhibits phorbol ester binding to protein kinase c via a competitive mechanism. **Biochem. Biophys. Res. Commun. 133** : 1051-1056.

Sharkey N.A., Leach K.L. & Blumberg P.M. (1984). Competitive inhibition by diacylglycerol of specific phorbol ester binding. **Proc. Natl. Acad. Sci.** (USA) **81** : 607-610.

Shearman M.S., Naor Z., Kikkawa U. & Nishizuka Y. (1987). Differential expression of multiple protein kinase c subspecies in rat central nervous tissue. **Biochem. Biophys. Res. Commun. 147** : 911-919.

Sigel E., Baur R. & Reuter H. (1988). Differential regulation by protein kinase c of foreign neuronal ion channels expressed in Xenopus oocytes. **Experientia 44**, A2.

Takai, Y., Kishimoto A., Inoue M. & Nishizuka Y. (1977). Studies on a cyclic nucleotide - independent protein kinase and its proenzyme in mammalian tissue. **J. Biol. Chem. 252** : 7603-7609.

Takai Y., Kishimoto A., Kikkawa U., Mori T. & Nishizuka Y. (1979). Unsaturated diacylglycerol as a possible messenger for the activation of calcium-activated, phospholipid-dependent protein kinase system. **Biochem. Biophys. Res. Commun. 91** : 1218-1224.

Tamaoki T., Nomoto H.,Takahashi I., Kato Y., Morimoto M. & Tomita F. (1986). Staurosporine, a potent inhibitor of phospholipid/Ca^{+2} dependent protein kinase. **Biochem. Biophys. Res. Commun. 135** : 397-402.

Tanaka C., Fujiwara H. & Fuji Y. (1986). Acetylcholine release from guinea pig caudate slices evoked by phorbol ester and calcium. **FEBS Lett. 195** : 129-134.

Tanaka C., Taniyama K. & Kusonocki M. (1984). A phorbol ester and A23187 act synergistically to release acetylcholine from the guinea pig ileum. **FEBS Lett. 175** : 165-169.

Watson S.P., McNally J., Shipman L.J. & Godfrey P.P. (1988). The action of the protein kinase c inhibitor, staurosporine, on human platelets ; evidence against a regulatory role for protein kinase c in the formation of inositol trisphosphate by thrombin. **Biochem. J. 249** : 345-350.

Witters L.A., Vater C.A. & Lienhard G.E. (1985). Phosphorylation of the glucose transporter in vitro and in vivo by protein kinase c. **Nature 315** : 777-778.

Wolf M., Cuatrecasas P. & Sayhoun N. (1985 a). Interaction of
protein kinase c with membranes is regulated by Ca^{+2},
phorbol esters, and ATP. **J. Biol. Chem.** 260 : 15718-15722.

Wolf M., Levine H., May W.S., Cuatrecasas P. & Sayhoun N.
(1985 b). A model for intracellular translocation of
protein kinase c involving synergism between Ca^{+2} and
phorbol esters. **Nature 317** : 546-549.

Woodgett J.R. & Hunter T. (1987). Immunological evidence for
two physiological forms of protein kinase c. **Mol. Cell.
Biol. 7** : 85-96.

Young S., Parker P.J., Ullrich A. & Stabel S. (1987). Down
regulation of protein kinase c is due to an increased rate
of degradation. **Biochem. J. 244** : 775-779.

BIOSYNTHESIS AND MODE OF INSERTION OF A STALKED INTRINSIC MEMBRANE PROTEIN OF THE SMALL INTESTINAL BRUSH BORDER

G. SEMENZA

Introduction

As revealed by electron microscopy (negative staining) the outer, luminal surface of brush border membranes is studied with lollipop-like proteins. These proteins have been identified with a number of enzymes, nearly all of them hydrolases. These proteins protrude into the lumen, and are attached to the membrane bilayer via a hydrophobic "stalk" and an embedded hydrophobic segment (hence the name of "stalked intrinsic membrane proteins" (Brunner et al., 1979) (see more below). As these enzymes of the plasma membrane exert their function on the <u>outside</u> of the membrane they are referred to as ectoenzymes. An extended but certainly not exhaustive list of the hydrolases of brush border membranes is given in Table 1 (for recent reviews see Kenny and Turner, 1987). Much attention has been paid in the last ten years or so to the mode of membrane insertion and biosynthesis of these proteins (for a recent review, see Semenza 1986).

The reason for this interest, and for our own interest in this biological object is manyfold : A) Stalked intrinsic membrane proteins, although far from occuring only in brush border membranes, are particularly abundant there, representing more thant 50 % of the intrinsic membrane proteins. Indeed, the synthesis of each of the major ectoenzymes in Table 1 may constitue 1 % of total cellular protein synthesis. Brush border ectoenzymes are therefore favorable objects for studying insertion and biosynthesis of plasma membrane proteins, and also of other classes of proteins (i.e., those which are inserted into or translocated across the ER membrane). B) With a single exception (as of today),

Table 1 : Some ectoenzymes of brush border membranes

A : Small-intestinal glycosidases (in most mammals)

A: Small-intestinal glycosidases (in most mammals)		EC No. (3.2.1.-)	Apparent molecular mass (kDa)	
			a	b
Lactase ⎫ Phlorizin-hydrolase ⎬ (glycosylceramidase) ⎭	The β-glycosidase complex	23 62, 45, 46	210	160
Trehalase		28	?	75
(Maltase)-glucoamylase 1 ⎫ (Maltase)-glucoamylase 2 ⎭	The glucoamylase complex	20	200	135 125
(Maltase)-sucrase ⎫ (Maltase)-isomaltase ⎭	The sucrase-isomaltase complex	48 10	200	120 140

a, of the non-glycosilated, primary translation product

b, of the mature form (or subunits)

Trehalase and glucoamylase are present in the renal brush border membranes also.

B : Proteases and peptidases in (pig) kidney brush border membranes

Class	Enzyme	Active site	Specificity	Subunit M_r ($\times 10^{-3}$)
Endopeptidases	Endopeptidase-24.11 (EC 3.4.24.11, ? all species)	Zn^{2+}	$-O-O-\overset{\downarrow}{\bullet}-O-$ (hydrophobic)	90
	Endopeptidase-2 (rat kidney)	Zn^{2+}	$-O-\overset{\downarrow}{\bullet}-O-O-$ (aromatic)	80, 74
	Meprin (mouse kidney)	Zn^{2+}	$-O-\overset{\downarrow}{\bullet}-O-O-$ (aromatic)	85
Aminopeptidases	Aminopeptidase N (EC 3.4.11.2)	Zn^{2+}	$\overset{\downarrow}{\bullet}-O-O-O-$ (many)	160
	Aminopeptidase A (EC 3.4.11.7)	Ca^{2+}	$\overset{\downarrow}{\bullet}-O-O-O-$ (Glu/Asp)	170
	Aminopeptidase W (EC 3.4.11.–)	Zn^{2+}	$O-\overset{\downarrow}{\bullet}-O-O-$ (Trp)	130
	Aminopeptidase P (EC 3.4.11.9)	?	$O-\overset{\downarrow}{\bullet}-O-O$ (Pro)	?
Carboxypeptidase	Carboxypeptidase P (EC 3.4.17.–)	Zn^{2+}	$-O-O-\overset{\downarrow}{\bullet}-O$ (Pro, Ala, Gly)	135
Dipeptidyl peptidase	Dipeptidyl peptidase IV (EC 3.4.14.5)	Serine	$O-\overset{\downarrow}{\bullet}-O-O-$ (Pro, Ala)	130
Peptidyl dipeptidase	Peptidyl dipeptidase A (ACE) (EC 3.4.15.1)	Zn^{2+}	$-O-O-\overset{\downarrow}{\bullet}-\bullet$ (nonspecific)	175
Dipeptidase	Microsomal dipeptidase (EC 3.4.13.11)	Zn^{2+}	$\overset{\downarrow}{\bullet}-\bullet$ (nonspecific)	50
Transferase	γ-Glutamyltransferase (EC 2.3.2.2)	–	$\overset{\downarrow}{\bullet}-O-O-O-$ (γ-Glu)	60, 30

Most of these proteases and peptidases have been identified in the small intestinal brush border membrane also. Other ectoenzymes of the brush border membranes include : alkaline phosphatase (EC 3.1.3.1), at least a phospholipase, etc. From Kenny and Turner, 1987.

the stalked intrinsic proteins of the brush border membranes seem to be anchored in one of two ways : via a hydrophobic stretch located in the N-terminal region of the polypeptide chain or via phosphatidylinositol. The hydrophobic stretches of the proteins of the former group appear therefore to be insensitive to signalase. Also, there is no known secretion from these cells into the lumen either of the intestine or the renal tubule. (Intestinal "secretion" is rather the result of membrane shedding or of the action of digestive proteases on the stalk of some stalked proteins). The question is why signalase does not cleave the signals of quite a few stalked ectoenzymes in this membrane. The significance of this problem reaches beyond brush border membranes.

To the group of stalked brush border proteins which are anchored via a hydrophobic segment located not far from the amino-terminal belong the sucrase-isomaltase complex (SI) (Fig. 1), the glucoamylase complex, γ-glutamyltransferase, endopeptidase 24.11, aminopeptidase N, dipeptidyl peptidase, and others. The complete sequences, deduced from cDNA cloning and sequencing, and hence the corresponding information on the mode of anchoring and presumed mechanism of biosynthesis and insertion, are available only for sucrase-isomaltase (Hunziker et al., 1986) γ-glutamyltransferase (Laperche et al., 1986), and endopeptidase 24.11 (Devault et al., 1987 ; Malfroy et al., 1987). Other brush border proteins which are likely to be anchored via an uncleaved signal and whose sequencing is underway include glucoamylase (Norén et al., 1986).

Originally, our own interest in intestinal disaccha-ridases was triggered by two observations which seemed to contradict one another, and thus puzzled our minds. Based on heat inactivation experiments, intestinal maltase-sucrase had been shown not to be identical with maltase-1,6-α-glucosidase

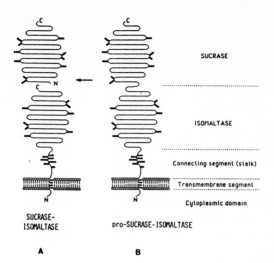

Fig. 1 : Positioning of sucrase-isomaltase (A) and of pro-sucrase-isomaltase (B) in the small-intestinal brush border membrane. The (unspecified) interactions within and between the sucrase and isomaltase domains (or subunits, respectively) are not indicated.
ooo, sugar chains. From Hunziker et al., 1986, modified.

```
(Met)Ala-Lys-Arg-Lys-Phe-Ser-Gly-Leu-Glu-Ile-Thr-Leu-Ile-Val-Leu-Phe-
     1            5                   10              15

Val-Ile-Val-Phe-Ile-Ile-Ala-Ile-Ala-Leu-Ile-Ala-Val-Leu-Ala-Thr-Lys-
         20              25              30

Thr-Pro-Ala-Val-Glu-Glu-Val-Asn-Pro-Ser-Ser-Ser-Thr-Pro-Thr-Thr-Thr-
    35              40              45              50

Ser-Thr-Thr-Thr-Ser-Thr-Ser-Gly-Ser-Val-Ser-Cys-Pro-Ser-Glu-
        55              60              65
```

Fig. 2 : N-Terminal sequence of rabbit pro-sucrase-isomaltase and of the isomaltase subunit of "ripe" sucrase-isomaltase complex. The hydrophobic stretch spanning the membrane (see Fig. 1) is underlined. Note also the Ser, Thr rich sequence beginning at pos. 43. From Frank et al., 1978 ; Sjöström et al., 1982 ; Hunziker et al., 1986.

(Dahlqvist, 1962) ; yet both maltase-sucrase and maltase-isomaltase were either absent or devoid of activity in sucrose-isomaltose malabsorption (e.g., Auricchio et al., 1965a) a
monofactorial genetic condition (e.g., Prader et al., 1961 ;
Kerry and Townley, 1965). Work by Auricchio, Kolinska, myself
and others (Auricchio et al., 1965b ; Kolinska and Semenza,
1967) began to sort out this apparent contradiction :
maltase-sucrase and maltase-isomaltase are the subunits of a
heterodimer, the "sucrase-isomaltase complex" ; although
similar in many respects (see below)they differ vastly in
their stability towards heat and other denaturing agents, so
that it is possible to inactivate in vitro one of the
subunits, leaving the activity of the other essentially
unaffected (Auricchio et al., 1965b ; Kolinska and Semenza,
1967 ; Cogoli et al., 1973).

In the subsequent years, work concentrated mainly in
two areas : the groups of Auricchio, Kretchmer, Koldovsky,
and others showed that sucrase and isomaltase (and, of
course, the related maltase activity) always developed simultaneously ; it was so in normal development, whether they
appear at the time of weaning (e.g., in the rat and mice) or
in intrauterine life (e.g., in man, reviewed by Mobassaleh et
al., 1985) ; it was also in the precocious development
triggered by corticosteroids, their response to diet and in
developing in vitro cultures of small intestines (for reviews
see Koldovsky 1981 ; Henning 1985 ; Kedinger et al., 1986).

Also, a large number of observations scattered throughout the biological and biochemical litterature indicated
that isomaltase activity is present in the small intestine of
all vertebrates investigated, while sucrase only occurs in
most (but not all) terrestrial mammals, birds and reptiles,
and seems to be absent in amphibia (reviewed by Semenza,
1968 ; 1981).

Parallel to this, knowledge was accumulated on the biochemical properties of the two subunits and on the mode of anchoring of the sucrase-isomaltase complex to the brush border membrane. To make a long story short : each subunit is composed of a single, glycosylated polypeptide chain (apparent mol. wt.-values, 120 kDa for sucrase, 140 kDa for isomaltase) ; a large number of their functional (and structural, when known) properties were either identical or very similar (e.g., the interaction of fully-competitive inhibitors, affinity labels, substrates with the active sites ; the kinetic and minimum catalytic mechanism ; the amino acid sequence of parts of the active sites, etc.) (Reviewed by Semenza, 1981 ; 1986).

The mode of anchoring of the sucrase-isomaltase complex to the brush border membrane proved to be unusual at that time : (Fig. 1) sucrase has a peripheral position and does not interact directly with the membrane bilayer (Brunner et al., 1979). Isomaltase, instead, is anchored to the brush border membrane solely via a highly hydrophobic segment (indeed, one of the most hydrophobic segments ever sequenced by classical means (Fig. 2), located in the N-terminal region of its polypeptide chain (Frank et al., 1978 ; Sjöström et al., 1982 ; Hunsiker et al., 1986). This mode of anchoring of isomaltase (and the similar one, soon and independently identified for two aminopeptidases (Macnair and Kenny, 1979 ; Louvard et al., 1976) was quite different from that of glycophorin, the only other plasma membrane protein, the anchoring of which was known in 1978 in some detail : the hydrophobic segment of glycophorin is located in its C-terminal region (Furthmayr et al., 1978) ; accordingly, the mode of biosynthesis and of cotranslational insertion of glycophorin proposed, which took into account this positioning, could not be applied to the sucrase-isomaltase complex.

Thus, in the late seventies a further group of challenging puzzles emerged : how can we explain the peripheral positioning of sucrase ? How were the current ideas of that time on synthesis and insertion of plasma membrane proteins to be modified in order to accomodate the anchoring of isomaltase via the N-terminal region ? Is there a way to explain in a single framework also the similar, and often identical properties of the sucrase and isomaltase subunits ; their similar, if not identical, biological control mechanism (as the biology of hemoglobin learns, the subunits of a heterodimer need not always be subjected to the same control mechanism), and, possibly, also the distribution of intestinal sucrase and isomaltase activities among vertebrates ?

The pro-sucrase-isomaltase hypothesis

In 1977 and 1978 I proposed the "one-chain, two-active sites precursor hypothesis" (Semenza, 1978 ; 1979 a, b), which aimed at explaining in a single framework phylogenetic, ontogenetic, developmental, cell-biological and biochemical aspects. This hypothesis can be formulated as follows (see Fig. 3) : (i) sucrase and isomaltase have arisen phylogenetically by (partial) duplication of an original isomaltase-maltase gene (see below) ; (ii) this would have led first to a gene coding for a single polypeptide chain carrying two identical domains, each endowed with enzymatic activity : a "double isomaltase", (iii) subsequent mutation would have transformed one of these domains from an isomaltase-maltase into a sucrase-maltase which resulted in an (active ?) single chain two active sites precursor ("pro-sucrase-isomaltase") ; (iv) this single-chain-pro-sucrase-isomaltase would be synthesized, glycosylated and inserted in the membrane of the endoplasmic reticulum and transferred, along with other plasma membrane proteins, to the brush border membrane ; (v) post-translational proteolytic modification of this single chain (perhaps by one or more pancreatic proteases), would lead to the two subunits of the "ripe" SI complex ; they would still remain associated via interactions formed during the folding of single-chain pro-sucrase-isomaltase. Note that we postulated the biosynthesis of a very unusually long polypeptide chain (of at least 260 kDa), i.e., of a nearly unprecendented size ; and that this chain would be split post-translationnally, an event for which no obvious biological advantage could be visualized (and for which still now we can only provide more or less reasonable guesses). Yet, the "pro-sucrase-isomaltase hypothesis", in addition to being correct has provided the leading thread to our understanding of much of the biology and biochemistry not only of sucrase-isomaltase, but also of other stalked brush border proteins.

Whenever sucrase-isomaltase develops without being exposed to pancreatic juice, it is always in the form of a single, gigantic pro-sucrase-isomaltase : in hog, whose duodenum had been disconnected from pancreas (Sjöström et al., 1980) ; in transplants of rat fetal small intestine under the skin of adult rats (Hauri et al., 1980 ; 1982 ; Montgomery et al., 1981) ; in rats, whose pancreatic ducts had been bypassed (Riby and Kretchmer, 1985) ; in human fetuses prior to the development of a functional pancreas (Skovbjerg, 1982) ; in intracellular membranes, during the homing process towards the brush border membrane (in rats, in vivo) (Hauri et al., 1979) ; same, in in vitro tissue cultures (Danielsen 1982 ; Danielsen and Cowell, 1985) ; same, in in vitro cell cultures (from human colonic cancer, Caco 2 (Hauri et al., 1985) ; small amounts of unprocessed pro-sucrase-isomaltase are found in most preparations of "mature" sucrase-isomaltase complex. Cell-free translation of intestinal mRNA leads to a very large (approx. 260 kDa), single-chain, immunoprecipitable pro-sucrase-isomaltase (Wacker et al., 1981). Finally, even the complete amino acid sequence of pro-sucrase-isomaltase is now available for cDNA cloning and sequencing (Hunziker et al., 1986). (For reviews, see Semenza, 1986 ; Danielsen et al., 1984 a).

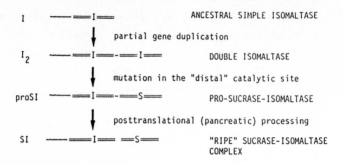

ANCESTRAL SIMPLE ISOMALTASE diagram

I   ━━━━━ ═══I═══   ANCESTRAL SIMPLE ISOMALTASE

↓   partial gene duplication

I_2 ━━━ ═══I═══-═══I═══   DOUBLE ISOMALTASE

↓   mutation in the "distal" catalytic site

proSI   ━━━ ═══I═══-═══S═══   PRO-SUCRASE-ISOMALTASE

↓   posttranslational (pancreatic) processing

SI   ━━━━ ═══I═══ ═══S═══   "RIPE" SUCRASE-ISOMALTASE
                                                 COMPLEX

Fig. 3 : The "One chain-two active sites precursor hypothesis" (Semenza, 1978, 1979 a, b).
〰〰〰〰 , the N-terminal region of the polypeptide chain, which interacts with the lipid bilayer. I, isomaltase. S, sucrase (both I and S have strong maltase activity).

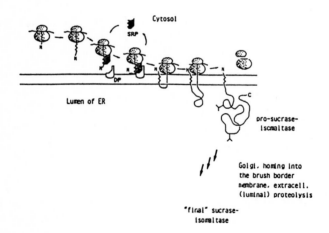

Fig. 4 : Proposed dual role of the hydrophobic sequence in the N-terminal region of pro-sucrase-isomaltase (see Fig. 2), as an uncleaved signal (in biosynthesis, this figure) and as the eventual anchor (Fig. 1) in the "final" protein. ER, endoplasmic reticulum. SRP, signal recognition particle. DP, docking protein (= SRP receptor). From Semenza, 1986, modif., based on Hunziker et al., 1986.

Points (iv) and (v) of the hypothesis mentioned above are thus demonstrated <u>ad abundatiam</u>. More difficult is of course, to demonstrate the existence of a "double isomaltase" in the distant evolutionistic past. However, an interesting single-chain two-active sites isomaltase was identified recently in the small-intestinal brush border membrane of the sea lion (<u>Zalophus californianus</u>). This enzyme is composed of a single polypeptide chain, carrying two apparently identical active sites, each splitting both maltose and isomaltose : it thus mimicks the postulated ancestral "double isomaltase" (Wacker et al., 1984).

Can we make reasonable guesses as to the time when the ancestral "double isomaltase" mutated into pro-sucrase-isomaltase ? (i.e., as to when sucrase activity appeared in evolution). A study on the properties and membrane insertion of chicken small-intestinal sucrase-isomaltase shows it to be anchored to the membrane via the isomaltase subunit and to arise via pro-sucrase-isomaltase (Hu et al., 1987). Thus, no substantial difference seems to exist in these respects between avian and mammalian sucrase-isomaltase, which tentatively places the mutation of double isomaltase to pro-su-crase-isomaltase (point (iii) above) prior to the separation of mammals from reptiles, i.e., to more than 330 Mio. years ago.

Point (i, ii) in the hypothesis above, i.e., the partial gene duplication giving rise to "double isomaltase", was suggested by the similarities between the subunits, which indicated extensive homology. The complete amino acid sequence of pro-sucrase-isomaltase via cDNA cloning and sequencing (Hunziker et al., 1986) has now confirmed it fully. As predicted, and as expected from previous work on the mode of anchoring of pro-sucrase-isomaltase and of sucrase-isomaltase the anchoring and the "stalk" segments are confined to the N-terminal portion ; no sequence(s) homologous to them are

found anywhere else. But contrary to this, and very remarkably, the isomaltase and sucrase portions of pro-sucrase-isomaltase show a highly significant degree of homologies and
analogy (approx. 41 % identical residues, 40 % conservative
changes) : It is beyond doubt, therefore, that in the
phylogeny of sucrase-isomaltase a <u>partial</u> gene duplication
has occurred.

Synthesis and membrane insertion

Knowledge of the complete sequence of the 1827 amino
acid residues of pro-sucrase-isomaltase and of its cDNA has
produced new insights as to its mode of biosynthesis and
insertion into the membrane. We could partially correct and
add further details to the mode of anchoring suggested
earlier. In particular, it is now clear that the N-terminus
of isomaltase is located at the cytosolic face of the
membrane ; the highly hydrophobic sequence beginning at pos.
12 in the mature protein spans the membrane once, and is
followed from pos. 32 on by a hydrophilic sequence, the most
characteristic feature of which is a high frequency of
Ser/Thr residues (17 out of 22, beginning at pos. 44), some
of which are glycosylated (and must, therefore, be at the
extracellular, luminal face of the membrane). This highly
hydrophilic stretch between the hydrophobic anchor and the
"body" of the protein (which is known to protrude as a
"lollipop" at the luminal side) is likely to be the "stalk".

From the cDNA sequence is also apparent that no
cleavable signal preceds the sequence of the final product,
pro-sucrase-isomaltase. In fact, in the cDNA-deduced sequence, Ala-2 (corresponding to Ala-1 of pro-sucrase-isomaltase)
is preceded by the initiation codon Met and only 9 bp
upstream by an in-frame stop codon. Furthermore, if no
cleavable signal sequence is synthesized prior to pro-sucrase-isomaltase proper, ^{3}H-leucine should be incorporated in

vitro (in a cell-free system, in the absence of microsomal membranes) at the same positions as it is found in the final, mature protein. This is indeed found to be the case (Ghersa et al., 1986). Clearly, the initiation Met only is cleaved off during biosynthetis and insertion of pro-sucrase-isomaltase. The only reasonable candidate to target this protein to the endoplasmic reticulum membrane during biosynthesis and insertion is the hydrophobic segment Leu-12 through Ala-31 (Fig. 2), which eventually spans the brush border membrane (Fig. 1). We thus proposed a dual role for this highly hydrophobic stretch, as it has already been proposed for neuraminidase and the asialoglycoprotein receptor (Spiess and Lodish, 1986). These proteins are positioned in the membrane on a way similar to that of pro-sucrase-isomaltase. A simple model for the process of membrane insertion of the nascent chain can thus be suggested, in which the chain forms a hairpin or loop as an antiparallel helical pair which gets engaged into the membrane bilayer (see e.g., Fig. 4).

From the cisternal side pro-sucrase-isomaltase is glycosylated to the "high-mannose-form", and than to the "complex form" (reviewed by Danielsen et al., 1984). Segregation and glycosylation have been recently demonstrated in a cell-free system also (Ghersa et al., 1986).

The existence of a common precursor in the biosynthesis of sucrase and isomaltase, i.e., of pro-sucrase-isomaltase, provides a straightforward explanation why sucrase and isomaltase respond in the same way to a number of stimuli - indeed it has become more interesting now to look for conditions which affect one activity independently of the other.

Sucrose-isomaltose malabsorption may prove to be a veryinteresting tool. In the various pedigrees of this condition different anomalies have been reported : existence

of an immunologically cross-reacting, enzymatically inactive
protein (Dubs et al., 1973), lack of such a protein (Schmitz
et al., 1973 ; Gray et al., 1976), existence of a large-M_r
isomaltase devoid of sucrase activity (Freiburghaus et al.,
1977), halt of pro-sucrase-isomaltase in the Golgi membranes,
etc. Indeed, considering the complex chain of events in the
ontogenesis of sucrase-isomaltase a variety of genotypes and
of phenotypes, all leading to sucrose-isomaltose malabsorption, are now conceivable.

Particularly interesting are of course the block(s)
in the route transferring pro-sucrase-isomaltase from ER to
the brush border membrane (Naim et al., 1988). They may well
provide insights in the mechanisms regulating the traffic of
membrane proteins in this route and perhaps also on the
mechanism(s) of sorting out.

**Pro-sucrase-isomaltase as a model of other stalked proteins
of the brush borders**

The pro-sucrase-isomaltase hypothesis has proven
fruitful not only in guiding investigations on the ontogenesis, phylogenesis and developmental control of sucrase-isomaltase, but also in suggesting similar events to take place
for other stalked proteins of the brush border membranes,
small-intestinal and renal.

In fact, the small-intestinal glucoamylase complex is
also synthetized as a large M_r pro-form, which is split by
pancreatic proteases ; also, its hydrophobic anchor is
located in an N-terminal region.

The β-glucosidase complex (lactase-glycosylcera-
midase) is likewise synthetized as a very large pro-form,
which is processed, however, intracellularly (Skovbjerg et
al., 1984 ; Danielsen et al., 1984 b ; Naim et al., 1987 ;
Büller et al., 1987).

The peptidases and proteases of the brush border membranes are not synthetized as large M_r "double" enzymes, but in some cases at least, their "anchor" is again localized in an N-terminal region. It is conceivable that they may act, at some point in the biosynthesis and membrane insertion processed, as non-cleaved "signals". Indeed, this has been demonstrated for γ-glutamyltransferase (Laperche et al., 1986) and for endopeptidase 24.11 (Devault et al., 1986 ; Malfroy et al., 1986).

As mentioned at the beginning, another group of brush border stalked proteins are anchored to the membrane fabric via phosphatidyl-inositol. This has been shown thus far for alkaline phosphatase (for a review, see Low, 1987), trehalase (Takesue et al., 1986) ; dipeptidase (Hooper et al., 1987). A comparison of their (cleavable) signal sequences with the uncleaved signals, and in general, of their biosynthetic mechanism will indicate how similar and how different the stalked intrinsic proteins of the brush border membranes are.

References

Auricchio S., Rubino A., Prader A., Rey J., Jos J., Frézal J. and Davidson M. (1965 a). Intestinal glucosidase activities in congenital malabsorption of disaccharidases. **J. Ped.** **66** : 555-564.

Auricchio S., Semenza G., Rubino A. (1965 b). Multiplicity of human intestinal disaccharidases. II. Characterization of the individual maltases. **Biochim. Biophys. Acta 96** : 498-507.

Brunner J., Hauser H., Braun H., Wilson K.J., Wacker H., O'Neill B., Semenza G. (1979). The mode of association of the enzyme complex sucrase-isomaltase with the intestinal brush-border membrane. **J. Biol. Chem. 254** : 1821-1828.

Büller H.A., Montgomery R.K., Sasak W.V., Grand R.J. (1987). Biosynthesis glycosylation, and intracellular transport of intestinal lactase-phlorizin-hydrolase in rat. J. Biol. Chem. **262** : 17206-17211.

Cogoli A., Eberle A., Sigrist H., Joss Ch., Robinson E., Mosimann H., Semenza G. (1973). Subunits of the small-intestinal sucrase-isomaltase complex and separation of its enzymatically active isomaltase moiety. **Eur. J. Biochem. 33** : 40-48.

Dahlqvist A. (1962). Specificity of the human intestinal disaccharidases and implications for hereditary disaccharide intolerance. **J. Clin. Invest. 41** : 463-470.

Danielsen E.M. (1982). Biosynthesis of intestinal microvillar proteins. Pulse chase labeling studies on aminopeptidase N and sucrase-isomaltase. **Biochem. J. 204** : 639-645.

Danielsen E.M., Cowell G.M. (1985). Biosynthesis of intestinal microvillar proteins. The intracellular transport of amino-peptidase N and sucrase-isomaltase occurs at different rates pre-Golgi but at the same rate past-Golgi. **FEBS Lett. 190** : 69-72.

Danielsen E.M., Cowell G.M., Norén, Sjöström H. (1984 a). Biosynthesis of microvillar proteins. **Biochem. J. 221** : 1-14.

Danielsen E.M., Skovbjerg H., Norén O., Sjöström H. (1984 b). Biosynthesis of intestinal microvillar proteins. Intracellular processing of lactase-phlorizin-hydrolase. **Biochem. Biophys. Res. Commun. 122** : 82-90.

Devault A., Lazure C., Nault Ch., Le Moual H., Seidah N.G., Chrétien M., Kahn Ph., Powell J., Mallet J., Beaumont A., Roques B.P., Crine P., Boileau G. (1987). Amino acid sequence of rabbit kidney neutral endopeptidase 24.11 (enkephalinase) deduced from a complementary DNA. **EMBO J. 6** : 1317-1322.

Dubs R., Steinmann B., Gitzelmann R. (1973). Demonstration of an inactive enzyme antigen in sucrase-isomaltase deficiency. **Helv. paediat. Acta 28** : 187-198.

Frank G., Brunner J., Hauser H., Wacker H., Semenza G., Zuber H. (1978). The hydrophobic anchor of small-intestinal sucrase-isomaltase. N-terminal sequence of the isomaltase subunit. **FEBS Lett. 96** : 183-188.

Freiburghaus A.U., Dubs R., Hadorn B., Gaze H., Hauri H.P., Gitzelmann R. (1978). The brush border membrane in hereditary sucrase-isomaltase deficiency : abnormal protein pattern and presence of immunoreactive enzyme. **Eur. J. Clin. Invest. 7** : 455-459.

Furthmayr H., Galardy R.E., Tomita M., Marchesi V.T. (1978). The intramembranous segment of human erythrocyte glycophorin A. **Arch. Biochem. Biophys. 185** : 21-29.

Ghersa P., Huber P., Semenza G., Wacker H. (1986). Cell-free synthesis, membrane integration, and glycosylation of pro-sucrase-isomaltase. **J. Biol. Chem. 261** : 7969-7974.

Gray G.M., Conklin K.A., Townley R.R.W. (1976). Sucrase-isomaltase deficiency. Absence of an inactive enzyme variant. **New Engl. J. Med. 14** : 750-753.

Hauri H.-P., Quaroni A., Isselbacher K. (1979). Biogenesis of intestinal plasma membrane : posttranslational route and cleavage of sucrase-isomaltase. **Proc. Natl. Acad. Sci. USA 76** : 5183-5186.

Hauri H.-P., Quaroni A., Isselbacher K.J. (1980). Monoclonal antibodies to sucrase-isomaltase : probes for the study of postnatal development and biogenesis of the intestinal microvillus membrane. **Proc. Natl. Acad. Sci. USA 77** : 6629-6633.

Hauri H.-P, Wacker H., Rickli E.E., Bigler-Meier B., Quaroni A., Semenza G. (1982). Biosynthesis of sucrase-isomaltase. Purification and NH$_2$-terminal amino acid sequence of the rat sucrase-isomaltase precursor (pro-sucrase-isomaltase) from fetal intestinal transplants. J. Biol. Chem. 257 : 4522-4528.

Hauri H.-P., Sterchi E.E., Bienz D., Fransen J.A.M., Marxer A. (1985). Expression and intracellular transport of microvillus membrane hydrolases in human epithelial cells. J. Cell Biol. 101 : 838-851.

Henning S.J. (1985). Ontogeny of enzymes in the small intestine. Ann. Rev. Physiol. 47 : 231-245.

Hooper N.M., Low M.G., Turner A.J. (1987). Renal dipeptidase is one of the membrane proteins released by phosphatidylinositol-specific phospholipase C. Biochem. J. 244 : 465-469.

Hu C., Spiess M., Semenza G. (1987). The mode of anchoring and precursur forms of sucrase-isomaltase and maltase-glucoamylase in chicken intestinal brush border membrane. Phylogenetic implications. Biochim. Biophys. Acta 896 : 275-286.

Hunziker W., Spiess M., Semenza G., Lodish H.F. (1986). The sucrase-isomaltase complex : Primary structure, membrane orientation and evolution of a stalked, intrinsic brush border protein. Cell 46 : 227-234.

Kedinger M., Haffen K., Simon-Asmann P. (1986). Control mechanism in the ontogenesis of villus cells. In Molecular and Cellular Basis of Digestion (Desnuelle P., Sjöström H., Norén O., eds.) Elsevier, Amsterdam, pp. 323-334.

Kenny A.J., Turner A.J. (1987). Mammalian Ectoenzymes. Elsevier Science Publishers B.V. (Biomedical Division).

Kerry K.R., Townley R.R. (1965). Genetic aspects of intestinal sucrase-isomaltase deficiency. Austr. Paediatr. J. 1 : 223-235.

Koldovsky O. (1981). Developmental, dietary and hormonal control of intestinal disaccharidases in mammals (including man). In : **Carbohydrate Metabolism and Its Disorders** (Randle J.P., Steiner D.F., Whelan W.J., eds) vol. 3 pp. 418-522. Academic Press, London.

Kolinska J., Semenza G. (1967). Studies on intestinal sucrase and on intestinal sugar transport. V. Isolation and properties of sucrase-isomaltase from rabbit small intestine. **Biochim. Biophys. Acta 146** : 181-195.

Laperche Y., Bulle F., Aissani T., Chobert M.N., Aggerbeck M., Hanoune J., Guellaen G. (1986). Molecular cloning and nucleotide sequence of rat kidney γ-glutamyltranspeptidase cDNA. **Proc. Natl. Acad. Sci. USA 83** : 937-941.

Louvard D., Sémériva M., Maroux S. (1976). The brush-border intestinal aminopeptidase, a transmembrane protein as probed by macromolecular photolabeling. **J. Mol. Biol. 106** : 1023-1035.

Low M.G. (1987). Biochemistry of the glycosyl-phosphatidylinositol membrane protein anchors. **Biochem. J. 244** : 1-13.

Macnair R.D., Kenny A.J. (1979). Proteins of the kidney microvillar membrane. The amphipathic form of dipeptidyl peptidase IV. **Biochem. J. 179** : 379-395.

Malfroy B., Schofield P.R., Kuang W.-J., Seeburg P.H., Mason A.J., Henzel W.J. (1987). Molecular cloning and amino acid sequence of rat enkephalinase. **Biochem. Biophys. Res. Commun. 144** : 59-66.

Mobassaleh M., Montgomery R.K., Biller J.A., Grand R.J. (1985). Development of carbohydrate absorption in the fetus and neonate. **Pediatrics 75 (suppl.)** : 160-165.

Montgomery R.K., Sybicki A.A., Forcier A.G., Grand R.J. (1981). Rat intestinal microvillus membrane sucrase-isomaltase is a single high molecular weight protein and fully active enzyme in the absence of luminal factors. **Biochim. Biophys. Acta 661** : 346-349.

Naim H.Y., Sterchi E.E., Lentze M.J. (1987). Biosynthesis and maturation of lactase-phlorizin hydrolase in the human small intestinal epithelial cells. **Biochem. J.** 241 : 427-434.

Naim H.Y., Roth J., Sterchi E., Lentze M., Milla P., Schmitz J., Hauri H.-P. (1988). Sucrase-isomaltase deficiency in man. Different mutations disrupt intracellular transport, processing and function of an intestinal brush border enzyme. **J. Clin. Invest.** in press.

Norén O., Sjöström H., Cowell G., Tranum-Jensen J., Hansen O.C., Welinder K.G. (1986). Pig intestinal microvillar maltase glucoamylase. Structure and membrane insertion. **J. Biol. Chem.** 261 : 12306-12309.

Prader A., Auricchio S., Mürset G. (1961). Durchfall infolge hereditären Mangels an intestinaler Saccharaseaktivität (Saccharose-intoleranz). **Schweiz. Med. Wochschr.** 91 : 465-468.

Riby J.E., Kretchmer N. (1985). Participation of pancreatic enzymes in the degradation of intestinal sucrase-isomaltase. **J. Paediatr. Gastroenterol. Nutr.** 4 : 971-979.

Schmitz J., Bresson J-L., Triadou N., Bataille J., Rey J. (1980). Analyse en électrophorèse sur gel de polyacrylamide des protéines de la membrane microvillositaire et d'une fraction cytoplasmique dans 8 cas d'intolérance congénitale au saccharose. **Gastroentérol. Clin. Biol.** 4 : 251-256.

Semenza G. (1968). Intestinal oligosaccharidases and disaccharidases. In : **Handbook of Physiology,** Sect. 6, Vol. V (C.F. Code, ed.) pp. 2543-2566. Washington D.C.

Semenza G. (1978). The sucrase-isomaltase complex, a large dimeric amphipathic protein from the small intestinal brush border membrane : Emerging structure-function relationships. In : **Structure and Dynamics of Chemistry.** (P. Ahlberg, L.-O., Sundelöf, eds.) Symp. 500th Jubilee, University of Uppsala, Sweden, 1977, pp. 226-240).

Semenza G. (1979 a). The mode of anchoring of sucrase-isomaltase to the small-intestinal brush border membrane and its biosynthetic implications. In : **Proc. 12th FEBS Meeting, Dresden 1978,** (S. Rapoport, T. Schewe, eds.) 53 : 21-28 Oxford/New York, Pergamon.

Semenza G. (1979 b). Mode of insertion of the sucrase-isomaltase complex in the intestinal brush border membrane : Implications for the biosynthesis of this stalked intrinsic membrane protein. In : **Development of Mammalian Absorptive Processes. Ciba Foundation Symposium** (K. Elliott, W.J. Whelan, eds.) 70 : 133-144, Amsterdam, Excerpta Medica.

Semenza G. (1981). Intestinal oligo- and disaccharidases. In : **Carbohydrate Metabolism and Its Disorders.** (P.J. Randle, D.F. Steiner, W.J. Whelan, eds.) 425-479, London, Academic Press.

Semenza G. (1986). Anchoring and biosynthesis of stalked brush border membrane proteins : Glycosidases and peptidases of enterocytes and renal tubuli. **Ann. Rev. Cell Biol. 2** : 255-313.

Sjöström H., Norén O., Christiansen L., Wacker H., Semenza G. (1980). A fully active, two-active site, single-chain-sucrase-isomaltase from pig small intestine. **J. Biol. Chem. 255** : 11332-11338.

Sjöström H., Norén O., Christiansen L.A., Wacker H., Spiess M., Bigler-Meier B., Rickli E.E., Semenza G. (1982). N-terminal sequences of pig intestinal sucrase-isomaltase and pro-sucrase-isomaltase. Implications for the biosynthesis and membrane insertion of pro-sucrase-isomaltase. **FEBS Lett. 148** : 321-325.

Skovbjerg H. (1982). High-molecular weight pro-sucrase-isomaltase in human fetal intestine. **Pediatr. Res. 16** : 948-949.

Skovbjerg H., Danielsen E.M., Norén O., Sjöström H. (1984). Evidence for biosynthesis of lactase-phlorizin-hydrolase as a single-chain high-molecular weight precursor. **Biochim. Biophys. Acta 798** : 247-251.

Spiess M., Lodish H.F. (1986). An internal signal sequence :
    The asialoglycoprotein receptor membrane anchor. **Cell 44** :
    177-185.
Takesue Y., Yokota K., Nishi Y., Taguchi R., Ikesawa H.
    (1986). Solubilization of trehalase from rabbit renal and
    intestinal brush-border membranes by a phosphoinositol-
    specific phospholipase C. **FEBS Lett. 201** : 5-8.
Wacker H., Jaussi R., Sonderegger P., Dokow M., Ghersa P.,
    Hauri H.-P., Christen Ph., Semenza G. (1981). Cell-free
    synthesis of the one-chain precursor of a major intrinsic
    protein complex of the small intestinal brush membrane
    (pro-sucrase-isomaltase). **FEBS Lett. 136** : 329-332.
Wacker H., Aggeler R., Kretchmer N., O'Neill B., Takesue Y.,
    Semenza G. (1984). A two-active site one-polypeptide
    enzyme : the isomaltase from sea lion small intestinal
    brush border membrane. **J. Biol. Chem. 259** : 4878-4884.

TOOLS AND APPROACHES TO STUDY MEMBRANES AND CELL ORGANELLES BIOGENESIS

N. LATRUFFE, A. KANTE, M. CHERKAOUI MALKI, A. BAILLY, P. ADAMI
& J.M. BERREZ

I. Introduction to diversity of cell compartments

Eucaryotic cells are highly compartimentalized ; indeed there are at least 10 to 12 membrane types forming different domains or microdomains (Palade, 1983). Cell membranes surround and limit distinct organelles which are either well individualized i.e. Chloroplasts, mitochondria or nucleus ; or organized as a reticulum i.e. endoplasmic reticulum (rough and smooth), golgi network ; or defined as vesicles with various functions i.e., endosomes, lysosomes, peroxisomes, or storage and secretion vesicles. In addition the plasma membrane limits the cell itself and isolates it from the external medium.

This high degree of structural organization allows a coordination and a very accurate control of cell metabolism and of the different cell functions and properties.

Biogenesis is a dynamic process which is vital for the cell. It is continuous and in equilibrium with the breakdown of macromolecules along of the cell growth or division. Biogenesis implies the addition of newly synthetized macromolecules into preexisting material (Warren, 1985) in order to preserve membrane selectivity and gradients (see further in § 5).

It is well kown that cell organelles and their subdomains have a specific composition as well towards their proteins identified as marker enzymes and marker antigens as, at a lesser degree, towards their lipids. However, this specific composition is not definitively fixed at the birth. There is a continuous and dynamic state including an intracellular traffic of proteins, of lipids and of membranes. Despite these flows which would lead to a randomization of all the

membranes compositions, the specificity remains remarkably
high. In addition we know that in some cell compartments there
are different specific domains. This is the case at least in
golgi, in lysosomes or in plasma membrane from polarized
cells.

Thereby the expression of the genetic program of
eucaryote cells must imply obligatory co-translational and
post-translational steps to permit newly synthetized polypeptides to have access to their final destination in a
functional state. In addition it is difficult to imagine the
existence of only one general mechanism allowing polypeptides
released from ribosomes to reach as many different destina-
tions as : lysosomes, mitochondria, plasma membrane, or other
compartments...

Biogenesis of cell compartments also concerns proca-
ryote organisms which own much more simple structures.
Meanwhile there is at least five different compartments in
gram⁻bacteria i.e. cytosol, plasma membrane, periplasm, outer
membrane and external space. Thus, neosynthetized bacterial
proteins must be also sorted (Pages, 1983). However in a
division cycle, bacteria need to synthetize membrane material
only 10 times less than animal cell since their membrane
surface requirement is 20 μm^2/30 min for bacteria and ≃ 10000
μm^2/24 h for eucaryote cells (Palade, 1983).

Attempts to elucidate the biosynthesis, the topogenesis and the dynamics of cell compartments are at the present
time one of the most exciting field of the contemporary
molecular cell biology. Although this research has been
initiated with the pioneer works of G. Palade in the fifties
(see Palade's Nobel lecture in 1975) with the discovery of
ribosomes and the exocrine pancreas properties, it really
opened out only fifteen years ago under G. Blobel and
coworkers'impulse resulting in the signal hypothesis in 1975
(Blobel & Dobberstein, 1975). In fact, the translocation of
proteins with often hydrophilic properties through largely
hydrophobic intracellular membranes must require different

signals, specific receptors and several structural modifications of the newly synthetized polypeptides ; for instance controlled proteolysis, glycosylation and other chemical modifications or conformational changes.

Several fundamental questions can be stated : How to explain the diversity of cell membranes ? How are the compartments organized ? How do they work ? How are the components sorted ? What is the molecular basis of the membrane specificity maintenance despite an active vesicular traffic which should lead to a randomization ? Finally which are the vectorial mechanisms involved in the displacement of proteins from a compartment to another ?

In order to solve these essential problems and thus, to fill up the hole between morphological observations and molecular mechanisms analysis, numerous and complementary techniques have been used, especially the tools of the molecular biology.

This paper is neither a review of this field nor an original work. Indeed, numerous good reviews already appeared in the past few years (Blobel, 1983 ; Palade, 1983 ; Zimmermann & Meyer, 1986 ; Loward, 1988) in almost each aspect of organelles biogenesis. But the spirit of this present article deals with the methodology and the pedagogy as it was required for the FEBS-CNRS 1987 advanced course on "Dynamics of membrane proteins and cellular energetics". Thus, this work has been written for the sake of mostly non familiar scientists with this field.

After the fundamental questions (see above), we present the main tools used in the subject and report the state of the knowledge at the beginning of 1988 by choosing two aspects that we develop in our team i.e. biogenesis of mitochondria and of peroxisomes.

2. Tools

The biogenesis of membranes and organelles is related to a wide domain of cell biology since it deals with the biosynthesis, the sorting and the assembly of their components which are essentially proteins/glycoproteins and lipids. To simplify, one can consider this field in terms of gene expression. The methodological approach starts at translational level of mRNA and implies to study the co-translational, post-translational, translocational and post-translocational events of the newly synthetized proteins.

These steps are required to deliver a protein at its right destination in a functional manner.

Primarily, basic evidences have often been obtained from appropriated biological models, for instance, cells with high protein synthesis activity or secretion i.e. exocrine pancreas (Palade, 1975), myeloma cells secreting immunoglobulins (Milstein et al., 1972). Yeasts i.e. Saccharomyces are also good material for their fast growth and division and the availability of many mutants. Decisive advances have been obtained with virus proliferation in animal cells i.e. vesicular stomatitis virus (VSV) (Darnell et al., 1986). Studies of molecular basis of some animal and human genetic diseases have been also very useful. Of course, bacteria, especially Escherichia coli, have been a choice material for the same reasons as yeasts. In these cases above cited, single molecules could be obtained in high amount and in a purified state and could be used as model for biogenesis studies.

Now, the current technology and the basic knowledges open the way to study more specialized systems or to elucidate the route of protein in minutes amounts : for example nervous cells and neuropeptides.

Organelle biogenesis is probably one of the biological fields which requires competence in the most numerous and various techniques of biochemistry, molecular and cell biology. First, classical techniques in isolation and characterization of proteins and mRNA are needed (Walsh et al., 1981).

Centrifugation methods are required for high purification of the different organelles and their subfractionation into individual microdomains (De Duve's Nobel lecture, 1975).

"In vitro" techniques like "in vitro" translation using various cell free systems : reticulocytes lysate, wheat germs lysate, yeast lysate or cell sap, are able to give significant advances in mechanism elucidation.

"In vitro" translocation and import tests can be developped by using microsomes or other organelles i.e. mitochondria, chloroplasts and peroxisomes...

And now "in vitro" glycosylation system, which allows to analyse the important post-translocational modifications of protein, will be soon available by using several microsomes or golgi vesicles.

All these reconstitution strategies cannot be operating without using antibody as probe of the studied protein. Thus, monoclonal or polyclonal antibodies and all the related immunochemical techniques are the necessary tools in this field (Delaage & Morel, 1988, this book).

With the development of the powerful genetic engineering technology in the eighties this subject is advancing with giant strides. Indeed cloned cDNA are more and more commonly used in transcription-translation "in vitro" system in order to synthetize substantial amount of precursors proteins. Recombinant DNA techniques and genes fusion can be use to construct any kind of chimaeric protein owning at will precise sequences of polypeptides coming from the site directed mutagenesis technology (Garoff, 1985).

Protein chemistry, peptide synthesis and physical techniques are also useful. One can complete with the general methods of cell biology, which are still appreciated including the electron microscopy technique (transmission, scanning,

freeze etching and freeze fracture) associated to autoradio-
graphy, and also the photonic microscopy for immunofluorescence. One must remain that these microscope techniques have been
the support of the first discoveries of this field about
thirty years ago. Now these conventional methods are more and
more sensitive and resolutive. They are usually associated
with imagery and computer programs.

These cellular approaches are particularly useful to
study membrane and vesicular traffic involved in endocyto-
sis-exocytosis processes (Klein et al., 1988, this book).

3. Approaches of membranes and organelles biogenesis

This problem can be dissociated into several aspects :
synthesis and delivery of proteins into their proper compartment and sorting mechanisms. The route endoplasmic reticulum
includes the synthesis of glycoproteins and their adressing,
the decoding mechanisms to differenciate the way of membrane
proteins from the luminal or the secreted ones. This implies
very accurate topogenesis events. Membranes and organelles
biogenesis concerns first the translocation of each protein
through at least one membrane whatever their final destination
and eventually involves active vesicular traffic.

Finally lipid biosynthesis and lipid flow are important activities in the cell biogenesis.

The last part of this topic which concerns import of
proteins into semi-autonomous organelles, i.e. chloroplast,
mitochondria and peroxisomes, is reported later.

A. The rough endoplasmic reticulum route of proteins

We must keep on mind that almost all the proteins
(98 %) are synthetized in the cytosol, while the last 2 % are
made into chloroplasts or mitochondria (Palade, 1983). According to this, one can estimate at least four different sorting
mechanisms for the cytoplasmically made proteins.

1. First sorting mechanism : the signal sequences

Proteins owning a signal sequence will not remain in the cytosol. The signal sequence (or leader sequence) is often a presequence located on the N.terminus side. Historically the first evidence for a precursor protein has been the light chain of an immunoglobulin (Milstein et al., 1972). Then Blobel and coworkers have generalized this property to several other proteins and proposed in 1975 the "signal hypothesis" (Blobel & Dobberstein, 1975). Now we know that many other proteins (luminal, secreted or membrane proteins) do not exhibit a presequence but an internal signal sequence which is not cleaved through the protein processing ; ovalbumine for instance. Although these signal sequences are not identical neither in length (15 to 70 amino-acids) nor in composition, they exhibit common physico-chemical properties i.e., a majority of hydrophobic or uncharged amino-acide residues, few basic residues, some or none hydroxylated residues, but almost never acidic one. Today, there is a general agreement to say that the amino-acid composition of a presequence is not the main signalling criterion while its spatial structure (mostly α-helix) would carry out the information. Important proofs have been brought by measuring interaction between synthetic signal sequence with phospholipid membranes (Allison & Schatz, 1986 ; Roise & Schatz, 1988) and by using chimaeric proteins obtained by genetic engineering (Lingappa et al., 1984). We know now that several signal sequences can exist into a protein (Sakaguchi et al., 1987). Similar signal sequences (or passenger sequences) are found in protein precursors destined to be translocated across bacterial plasma membrane or imported into chloroplasts or mitochondria (Roise et Schatz, 1988).

Translocation across the rough endoplasmic reticulum (R.E.R.)

        From several works (Blobel, Dobberstein, Meyer, Saba-
tini, Walter, see ref.), it is now well established that the
building polypeptide chain will be oriented to the endoplasmic
reticulum (ER) owing to the involvement of several cell
compounds, especially the signal recognition particle (SRP)
(Gilmore et al., 1982 ; Gilmore & Blobel, 1983) which is in
equilibrium between ER membrane, cytosol, ribosomes and the
SPR - receptor (or so called docking protein (DP)).
        The SRP (Walter & Blobel, 1981) is a ribonucleopar-
ticle containing a 7S RNA and 6 polypeptides, total MW of
250 kd. It is able to link the nascent polypeptide chain ($\simeq$ 70
amino-acids) and to arrest the cytosolic translation of
proteins exhibiting a signal sequence. The DP is an integral
protein of the RER with a molecular weight of 70 kd and plays
a receptor role for the preformed complex of polysomes - SPR
(Meyer et al., 1982). This last complex is also linked to RER
by two membrane proteins, the ribophorin I and II (Sabatini et
al., 1982).
        At this step the translation can start again. The
neosynthetized protein can be translocated through the membrane owing to its signal sequence (or insertion sequence) and
to a membrane channel (or pore) (Walter et al., 1984, 1986).
This process, which is a co-translational translocation, will
differ according to the type of protein.

        2. Second sorting mechanism : the stop-transfer
sequences

        Integral membrane proteins destined to stay in the RER
membrane or to migrate by lateral diffusion and by vesicular
traffic in other cell membranes, i.e. smooth endoplasmic
reticulum, golgi, lysosomes or plasma membrane, will not be
totally translocated across RER membrane (Vaz et al., 1984).
This mechanism is due, in addition to the insertion sequence,

to one or several stop transfer sequences with hydrophobic properties allowing interaction with phospholipid core (Mize et al., 1986).

In contrast, luminal or secreted proteins do not exhibit any stop transfer sequences and cannot remain anchored to the membrane.

Concerning the driving force of the protein translocation, one thought for a long time that the only energy was coming from the protein synthesis machinery. However recent reports demonstrate the direct ATP requirement for translocation (Hansen et al., 1986 ; Weich et al., 1987). The translocation process across RER is similar to the other membrane translocation mechanisms, i.e. export across bacteria plasma membrane (Pages, 1983), import into inner chloroplast membrane (Schmidt & Mishkind, 1986), into inner mitochondrial membrane (Hurt & Van Loon, 1986) or into peroxisomal membrane (Lazarow, 1985) but, in addition to ATP hydrolysis, such systems need also a membrane electrico potential (Eilers et al., 1987).

After (or during) the late translocation step there is often a large variety of co- and post-translational modifications which create vectorial and irreversible processes.

3. Third sorting mechanism : maturation, glycosylation and other post-translational modifications

If the signal presequence is present in a precursor, it will be cleaved by a luminal side oriented membrane signal peptidase according to a co-translational process. Conformational change of the protein is also a common mechanism.

Glycosylation (N and O linked) is the chemical modification which concerns most of the proteins whatever luminal, secreted or membrane bound they are (Farquahr, 1985). One knows that the primary N-glycosylation starts in the RER by the transfer of the same block of oligosacharides, containing 14 glycan residues, from the dolichol pyrophosphate (a membrane lipid) to one or several asparagine residue(s) located in a sequence of asn-x-thr (Kornfeld & Kornfeld, 1985). This step transfer is rapidly followed by a trimming which is initiated in the RER and continued in the cis-golgi. The final glycosylation is specific for each protein family and involves the coordinated action of glycosydases and glycosyltransferases which are located in the golgi network (cis, median and trans golgi vesicles). Protein transfer from RER to golgi is mediated by a vesicular transport which implies budding, fission and membranes fusion. The specific glycan residues composition of glycoproteins is now considered as a zip code. Presently the adressing of lysosomal proteins is the only best known mechanism. The tag is a mannose 6-phosphate residue (Von Figura & Hasilik, 1986). In this mechanism, lysosomal enzymes are recognized by mannose 6-phosphate receptor located either in cis-golgi vesicles or at the surface of the plasma membrane. The enzyme-receptor complexes are targeted into lysosomal vesicles where the receptor can be recycled (Genze et al., 1985 ; Klein et al., 1988).

There are numerous other post-translational modifications. For instance : - the sulfation of tyrosyl residues in the golgi apparatus which concerns most of the secreted proteins (Huttner, 1987) ; - the acylation (mainly myrystylation (Schmidt, 1982)) which allows hydrophilic proteins to be anchored into membranes ; - disulfide bonds formation ; - phosphorylations ; - protein folding (Rothman & Kornberg, 1986) ; - oligomeric association ; - addition of cofactors or prosthetic groups (Brambl & Plesofsky-Vig, 1986).

4. Fourth sorting mechanism : the vesicular traffic

The absence of physical continuity between organelles is only apparent. The linkage is ensured by an intense vesicular traffic which continuously transports molecules from a compartment to another allowing a spatial, temporal and

functional continuity. One could think that this vesicular
flow would randomize membrane and organelle composition. In
fact it is absolutely not the case, this flow is controlled by
a large variety of not yet well understood mechanisms
including ligand-receptor couples and other intervening molecules : clathrin of coated vesicles, cytosckeletal, microfilaments, intermediary filament, microtubules. For instance the
kinesin with ATPase activity is directly involved in the
organelle movement through microtubules (Vale et al., 1986 ;
Vale, 1987). Metabolic factors have also an important role in
the flow and in the communication between vesicles (i.e. pH,
H^+-ATPase...). Endocytosis and exocytosis processes which are
the main aspects of the vesicular traffic have been detailed
elsewhere (Baggiolini et al., 1988 ; Glick and Rothman, 1987 ;
Goldstein et al., 1985 ; Klein et al., 1988 ; Misek et al.,
1984 ; Scheckman, 1985 ; Steinmen et al., 1983).

B. Post-translational insertion of membrane proteins

Other mechanisms than the RER route have been demonstrated. They concern of course the proteins imported into
organelles (see further § 4) but also numerous proteins
completely synthetized on cytosolic free polysomes, then
discharged in the cytosol and finally spontaneously, specifically or not, inserted into membrane lipid bilayer (Sakaguchi
et al., 1987). For instance, cytochrome b_5 inserts at the
cytosolic face of endoplasmic reticulum while NADH-cytochrome
b_5 reductase and NADPH-cytochrome P450 reductase exhibit a
large distribution : cytoplasmic side of ER, golgi, plasma
membrane, outer mitochondrial membrane of adrenal cortex. This
type of membrane protein does not show any signal presequence.
The insertion mechanism in the lipid bilayer agrees with the
Wickner hypothesis of "Membrane triggered folding mechanism"
(Wickner & Lodish, 1983), where the conformation of the
precursor protein changes in the proximity of membrane
phospholipids and exposes their hydrophobic sequences. Other

examples have been reported with procaryotes i.e. the M13 coat
protein (Wickner, 1983), porin (Kleffel et al., 1985) or the
protein lam B (Charbit et al., 1986). The role of phospholipids in the proteins transfer from the cytosol to a membrane
is more and more suggested (Rietweld & De Kruijff, 1986).
Indeed the formation of protein-lipid micelles should largely
make easier the proteins insertion into membranes.

The insertion mechanism of membrane proteins must keep
the asymetric assembly of components. Indeed, as estabished by
Blobel (1983), the protein insertion can be monotopic, bitopic
or multitopic with respect to the different orientation of N
and C termini (Semenza, 1988 ; Singer et al., 1987 ; Yost et
al., 1983).

C. Lipid flow

Phospholipids, cholesterol and complex lipids are
essential components of biomembranes. It is well known they
have an asymetric distribution through the two layers (Williamson et al., 1987). Most of the enzymes involved in the
lipid biosynthesis are located on the cytosolic side of
endoplasmic reticulum (Bishop & Bell, 1985 ; Kawashima & Bell,
1987). However some of them are present in other organelles :
in mitochondria for disphosphatidylglycerol and phosphatidic
acid synthesis ; in chloroplasts which are able to synthetize
most of their lipids especially galactolipids ; and in
peroxisomes for the synthesis of plasmalogens.

Due to their physico-chemical properties and their
rapid lateral diffusion, lipids are easily transported from ER
to other types of membrane by the vesicular traffic. Lipid
transfer from ER to mitochondria is carried out by the
cytosolic phospholipid transfer protein (Wirtz, 1974 ; Bozzato
& Tinker, 1987). Micellar lipid transport seems to occur also.
So far, mechanisms involved in the lipid flow are not well
understood (Pagano & Sleight, 1985 ; Scow & Blanchette-Hackie,
1985). Membrane lipid asymetry is created and maintained by

membrane translocases. For instance the flipase of erythrocy
tes has been recently demonstrated (Zachowski et al., 1987).
Although lipids are known to be not chemically
reactive and not involved in specific functions except for
bioactive lipids in membrane signalling mechanisms i.e.
phosphoinositides in protein kinase c (Mahoney & Azzi, 1988,
this book), PAF acether in platelet activation (Roubin et al.,
1983) or phosphatidylcholine in D-β-hydroxybutyrate dehydrogenase (Latruffe et al., 1986). Membrane lipids play an
important role in organelle biogenesis. Indeed, due to their
fluidity and to their organization either as bilayer, mono-
layer, micelles, hexagonal-structure or as monomer, they
actively participate in the insertion and in the translocation
of protein through membranes and in the vesicular transport
(fusion, fission or division of organelles).

**4. Import of proteins into semi-autonomous organelles (chloro-
plasts and mitochondria) and into microbodies (peroxisomes and
glyoxisomes)**

Biogenesis of these organelles does not follow the RER
route since several polypeptides, destined to inner membrane
or thylakoids, are synthetized into mitochondria and chloroplasts (not discussed here) while most of them are made on
cytosolic free polysomes and post-translationally imported
into these organelles (see figure 1) (Colman & Robinson,
1986 ; Douglas et al., 1986). Mitochondrial DNA and, in a
lesser extent, chloroplast DNA are known in terms of nucleotide sequence and genes identification (Tzagoloff & Myers,
1986). But the post-translational insertion mechanisms of
proteins coded in these organelles are almost unknown (Clayton, 1984).

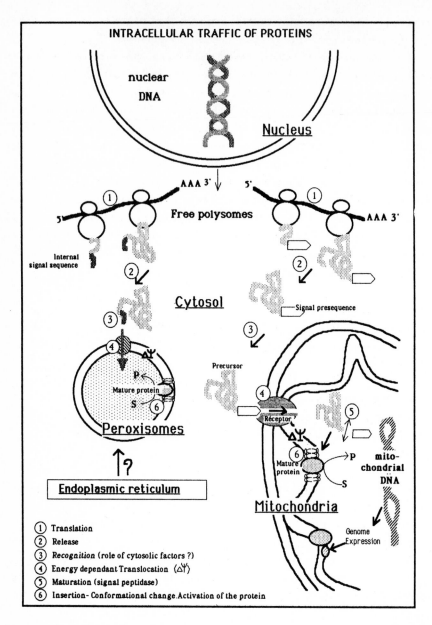

Figure 1 : Proteins import into mitochondria and peroxisomes

Two fundamental questions can be stated.

1. How can cytoplasmically made proteins be sorted in order either to remain into cytosol, to be targeted into mitochondria (or chloroplasts), to be targeted into peroxisomes (or glyoxisomes), or to be spontaneously inserted into other membranes ? (see above 3 § B) (Hurt & Van Loon, 1986).

2. How are imported proteins sorted into the organelle i.e. to their subcompartments which are at least four in mitochondria : outer membrane, inner membrane, inter membrane space and matrix (Van Loon et al., 1986), and also four in chloroplasts : outer membrane, inner membrane + thylakoïds, inter membrane space and stroma (Schmidt & Mishkind, 1986).

A. Mitochondria/chloroplasts

Numerous results appeared for the last ten years particularly from Schatz and Neupert laboratories (see references). Most of the proteins imported into mitochondria are synthetized under a larger precursor (all of them when destined for matrix). They own a N-terminus side presequence (called passenger sequence) which has the same properties as a signal sequence involved in the RER route (Roise & Schatz, 1988). Several proteins do not show any presequence i.e. all the proteins of the outer membrane and some of the inner membrane ; for instance cytochrome c or ADP-ATP translocator.

It is now demonstrated that translocation through mitochondrial membranes occurs at contact sites between outer and inner membrane (Pfanner et al., 1987). These translocation sites appear to present hydrophilic structures involved in the proteins translocation (Schwaiger et al., 1987). Proteins translocation through inner membrane requires both ATP (or GTP) hydrolysis and membrane potential (Pfanner & Neupert, 1986) while outer membrane proteins do not. A Zn-matrix protease cleaves, if needed, the passenger sequence. For proteins located in the intermembrane space, for instance cytochrome b_2 and for the Rieske Fe/S protein of complex III (Hartl et al., 1986), it has been shown that the processing operates in two steps : translocation in the matrix and first proteolytic cleavage, and then a back translocation through inner membrane and second proteolytic processing by a protease located on the external surface of the inner membrane. Evidence for different classes of mitochondrial receptors involved in the recognition of cytoplasmically made precursors has been reported. Such mitochondrial receptors have not been isolated yet while it has just been identified for chloroplasts envelop (Pain et al., 1988).

Cytosolic factors not characterized yet are involved in the sorting of proteins destined for mitochondria (Joste et al., 1987 ; Kanté et al., 1988, this book).

Elegant and powerfull experiments of gene fusion which allows to prepare chimaeric proteins have demonstrated common mechanisms between import of proteins into mitochondria and chloroplasts and export of proteins across bacterial plasma membrane (Lee & Beckwith, 1986 ; Lingappa et al., 1984). Indeed recent works have reported that cytosolic proteins or foreign proteins of an organelle can be targeted into such an organelle if they are coupled with a passenger sequence at its N-terminus side (Emr et al., 1986 ; Freeman et al., 1986 ; Hurt et al., 1986). One exception has pointed out the possible role of carboxyl terminus in the import process (Ness & Weiss, 1987). In addition spectacular observation has been published by Schatz's Laboratory (Hurt & Schatz, 1987) where they showed that the cytosolic dihydrofolate reductase (DHFR) contains a cryptic sequence which is able, when properly located, to oriente this protein into mitochondria. An interesting study on properties of mitochondrial presequences appeared recently (Roise & Schatz, 1988).

B. Microbodies (peroxisomes/glyoxisomes)

Although it was still thought few years ago that peroxisomes should derive from RER by budding, the works of Lazarow and coworkers have given evidence that peroxisomal proteins (wathever soluble or membrane proteins) are imported post-translationally from cytosolic free polysomes (Lazarow, 1985). No larger precursors have been report excepted for malate dehydrogenase from plant glyoxisomes (Borst, 1986). The import in the organelle is an energy dependent process where membrane potential is the main energy source (Bellion & Goodman, 1987). So far neither any surface receptor nor cytosolic factor involvements have been reported. A peroxisomal porin could play a role in the translocation mechanism. Nothing is known concerning the membrane phospholipids assembly. Taking into account that there is no relationship between ER and peroxisomes, a phospholipid transfer protein could be involved in the phospholipid traffic.

As other cellular organelles, new mitochondria, new chloroplasts and new peroxisomes are made by adding new lipid and protein material into preexisting organelles followed of buddings or division. The signals of these last mechanisms are poorly understood (Warren, 1985).

C. Nucleus biogenesis

Biogenesis of nucleus and nuclear membranes is complex since this organelle can transciently disappear along the cell mitosis (at the prophase). At least two mechanisms are involved : assembly of proteins from the RER route since the outer nuclear membrane is in continuity with ER and post-translational import of protein at contacts sites (pore sites) between both inner and outer nuclear membranes (Dingwall & Laskey, 1986).

5. Remarks and unsolved questions

Current knowledge shows both the diversity and the complexity of sorting mechanisms and intracellular traffic but also a significant unicity from procaryotes to eucaryotes (lower eucaryotes, plants and animals) as spectacularly illustrated by cross constructions obtained by genetic engineering.

Related to this recent findings Deshaies et al. (1988) demonstrated that 70 kd cytosolic heat shock proteins from yeast stimulate both translocation of secretory and mitochondrial precursor polypepides.

According to the fact that there is never "do novo" synthesis of new cell compartments, one can postulate that each membrane functions as its own assembly template (Palade, 1983).

One of the unclear aspect concerns the maintenance of membrane specificity despite an active vesicular traffic. Cytoskeletal and other endocellular proteins appear to play a decisive role in this process.

The cell polarity and the membrane polarity are also poorly understood (Misek et al., 1984). Until now the lipid flow receives much less attention than the protein counterpart while the lipids probably play decisive role in cell biogenesis.

One can think that the proteins sequences, thus their gene sequences, contain many informations which must be decoded to allow the proteins sorting and the insertion into their proper organelle.

So far, only little is known about the regulation of the assembly of mitochondrial or chloroplasts protein complexes which are under the control of nuclear genes or coded by the organelle DNA.

Finally, at this date, almost no real attention has been focused on the breakdown regulation of organelle components and their relationships with biosynthetic processes. Further works on this direction should bring new insight in cell membranes and organelle biogenesis.

Acknowledgments

Thanks are due to all of our coworkers and external collaborations in the progress of this field in our lab and also to Dr Y. Gaudemer for his continuous encouragements.

References :

Allison D.S. & Schatz G. (1986). Artificial mitochondrial presequences. **Prot. Natl. Acad. Sci. USA 83** : 9011-9015.

Baggiolini M., Deranleau D.A., Dewald B., Thelen M., Von Tsharner V. & Wymann M.P. (1988). The neutrophil leucocyte properties and mechanism of activation. (this book).

Bellion E. & Goodman J.M. (1987). Proton ionophores prevent assembly of a peroxisomal protein. **Cell 48** : 165-173.

Bishop W.R. & Bell R.M. (1985). Assembly of the endoplasmic reticulum phospholipid bilayer : the phosphatidylcholine transporter. **Cell 42** : 51-60.

Blobel G. (1983). Control of intracellular protein traffic, in "**Methods Enzymol.**". Vol **96** pp 663-682 (S. Fleischer & B. Fleischer eds) Acad. Press.

Blobel G. & Dobberstein B. (1975). Transfer of proteins across membranes. II. Reconstitution of functional rough microsomes from heterologous components. **J. Cell Biol. 67** : 852-862.

Borst P. (1986). How proteins get into microbodies (peroxisomes, glyoxysomes, glycosomes). **Biochim. Biophys. Acta 866** : 179-203.

Bozzato R.P. & Tinker D.O. (1987). Purification and properties of two phospholipid transfer proteins from yeast. **Biochem. Cell Biol. 65** : 195-202 and 203-210.

Brambl R. & Plesofsky-Vig N. (1986). Panthotenate is required
   in Neurospora crassa for assembly of subunit peptides of
   cytochrome c oxidase and ATPase/ATPsynthase. **Proc. Natl.
   Acad. Sci. USA 83** : 3614-3648.

Charbit A., Boulain J.C., Ryter A. & Hofnung M. (1986).
   Probing the topology of a bacterial membrane protein by
   genetic insertion of a foreign epitope ; expression at the
   cell surface. **EMBO J. 5** : 3029-3037.

Clayton D.A. (1984). Transcription of the mammalian mitochondrial genome in "**Ann. Rev. Biochem.**" Vol. **53**, pp 573-594.

Colman A. & Robinson C. (1986). Protein import into organelles : hierachical targeting signals. **Cell 46** : 321-322.

Darnell J., Lodish H. & Baltimore D. (1986). **Molecular Cell
   Biology**. 1187 p . Scientific America Books inc. Freeman &
   Co. New York.

De Duve C. (1975). Exploring cells with a centrifuge. **Science
   189** : 186-194.

Delaage M. & Morel A. (1988). Monoclonal antibodies for
   immunoanalysis (this book).

Deshaies R.J., Koch B.D., Werner-Washburne M., Craig E.A. &
   Shekman R. (1988). A subfamily of stress proteins facilita-
   tes translocation of secretory and mitochondrial precursor
   polypeptides. **Nature 332** : 800-805.

Dingwall C. & Laskey K.A. (1986). Protein import into the cell
   nucleus in "**Ann. Rev. Cell Biol.**". Vol 2, pp 367-390 (G.
   Palade, B.M. Alberts & J.A. Spudich eds) Ann. Rev. Inc,
   Palo Alto CA.

Douglas M.G., Mc Cammon M.T. & Vassarotti A. (1986). Targeting
   proteins into mitochondria. **Microbiol. Rev. 50** : 166-178.

Eilers M., Oppliger W. & Schatz G. (1987). Both ATP and
   energized inner membrane are required to import a purified
   precursor protein into mitochondria. **EMBO J. 6** : 1073-1077.

Emr S.D., Vassarotti A., Garrett J., Geller B.L., Takeda M. & Douglas H.G. (1986). The amino terminus of the yeast F_1-ATPase β-subunit precursor functions as a mitochondrial import signal. J. Cell Biol. 102 : 523-533.

Farquahr M.G. (1985). Progress in unraveling pathways of golgi traffic in "Ann. Rev. Cell Biol.", Vol 1, pp 447-488 (G. Palade, B.M. Alberts & J.A. Spudich eds) Ann. Rev. Inc, Palo Alto CA.

Freeman K.B., Yatscoff R.N. & Ridley R.G. (1986). Experimental approaches to the study of the biogenesis of mammalian mitochondrial proteins. Biochem. Cell Biol. 64 : 1108-1114.

Garoff H. (1985). Using recombinant DNA technique to study protein targeting in the eucaryotic cell in "Ann. Rev. Cell Biol", Vol 1 pp 403-445 (G. Palade, B.M. Alberts & J.A. Spudich eds) Ann. Rev. Inc, Palo Alto CA.

Geuze H.J., Slot J.W., Strous G.J.A.M., Hasilik A. & Von Figura K. (1985). Possible pathways for lysosomal enzyme delivery. J. Cell Biol. 101 : 2253-2262.

Gilmore R., Blobel G. & Walter P. (1982). Protein translocation across the endoplasmic reticulum. I. Detection in the microsomal membrane of a receptor for the signal recognition particle. J. Cell Biol. 95 : 443-469.

Gilmore R. & Blobel G. (1983). Transient involvement of signal recognition particle and its receptor in the microsomal membrane prior to protein translocation. Cell 35 : 677-685.

Glick B.S. & Rothman J.E. (1987). Possible role for fatty acyl-coenzyme A in intracellular protein transport. Nature 326 : 309-312.

Goldstein J.L., Brown M.S., Anderson R.G.W., Russell D.W. & Schneider W.J. (1985). Receptor mediated endocytosis, in "Ann. Rev. Cell Biol.", Vol 1 pp 1-39 (G. Palade, B.M. Alberts & J.A. Spudich eds) Ann. Rev. Inc, Palo Alto CA.

Hansen W., Garcia P.D. & Walter P. (1986). In vitro protein translocation across the yeast endoplasmic reticulum : ATP-dependant post-translational translocation of the prepro-α-factor. Cell 44 : 801-812.

Hartl F.V., Schmidt B., Wachter E., Weiss H. & Neupert N. (1986). Transport into mitochondria and intracellular sorting of the Fe/S protein of ubiquinol-cytochrome c reductase. Cell 47 : 939-951.

Hurt E.C., Goldschmidt-Clémont M., Pesold-Hurt B., Rochaix J.D. & Schatz G. (1986). A mitochodnrialpresequence can transport a chloroplast encoded protein into yeast mito-chondria. J. Biol. Chem. 261 : 11440-11443.

Hurt E.C. & Schatz G. (1987). A cytosolic protein contains a cryptic mitochondrial targeting signal. Nature 325 : 499-503.

Hurt E.C. & Van Loon A.P.G.M. (1986). How proteins find mitochondria and intramitochondrial compartments. Trends Biochem. Sci. 11 : 204-207.

Huttner W.B. (1987). Protein tyrosine sulfation. Trends Biochem. Sci. 12 : 361-363.

Joste V., Berrez J.M., Latruffe N. & Nelson B.D. (1987). An import factor in reticulocyte lysates which stimulates processing of several precursors destinated for the rat liver mitochondrial inner membrane. Acta Chem. Scand. ser. B 41 : 770-772.

Kanté A., Berrez J.M. & Latruffe N. (1988). Synthesis and maturation of D-β-hydroxybutyrate dehydrogenase (BDH) from mitochondrial inner membrane (this book).

Kawashima Y. & Bell R.M. (1987). Assembly of the endoplasmic reticulum phospholipid bilayer. Transporters for phosphati-dylcholine and metabolites. J. Biol. Chem. 262 : 16495-16502.

Kleffel B., Garavito R.M., Baumeister W. & Rosenbush J.P. (1985). Secondary structure of a channel-forming protein : porin from E. coli outer membrane. EMBO J. 4 : 1589-1592.

Klein G., Bof M., Gonzalez C. & Satre M. (1988). Fluide-phase endocytosis and lysosomal enzyme excretion ; two facets of intracellular membrane traffic (this book).

Kornfeld R. & Kornfeld S. (1985). Assembly of asparagine linked oligosaccharides in "**Ann. Rev. Biochem.**" Vol 54, pp 631-664 (Snell, Boyer, Meister, Richardson eds) Ann. Rev. Inc, Palo Alto CA.

Latruffe N., Berrez J.M. & El Kebbaj M.S. (1986). Lipid-protein interactions in biomembranes studied through the phospholipid specificity of D-β-hydroxybutyrate dehydrogenase. **Biochimie 68** : 481-491.

Lazarow P.B. (1985). Biogenesis of peroxisomes, in "**Ann. Rev. Cell Biol.**", Vol **1**, pp 489-530. (G. Palade, B.M. Alberts & J.A. Spudich eds) Ann. Rev. Inc, Palo Alto CA.

Lee C. & Beckwith J. (1986). Co-translational and post-translational protein translocation in procaryotic systems ; in "**Ann. Rev. Cell Biol.**" Vol 2, pp 315-336 (G. Palade, B.M. Alberts & J.A. Spudich eds) Ann. Rev. Inc ; Palo Alto CA.

Lingappa V.R., Chaidez J., Yost C.S. & Hedgpeth J. (1984). Determinants for protein localization β-lactamase signal sequence directs globin across microsomal membranes. **Proc. Natl. Acad. Sci. USA 81** : 456-460.

Louvard D. (1988). Trafic et signalisation intracellulaire in "**Communication cellulaire et pathologie**" pp 114-120 (C. Kordon & L. Degos eds) INSERM John Libbey eurotext, London, Paris.

Mahoney C.W. & Azzi A. (1988). Structure and function of Ca^{++} and phospholipid dependent protein kinase (protein kinase c), a tranmembrane signal transducer (this book).

Meyer D.I., Krause E. & Dobberstein B. (1982). Secretory protein translocation across membranes ; the role of "docking protein" **Nature 297** : 647-650.

Milstein C., Brownlee G., Harrison T., Mathews M.B. (1972). A possible precursor of immunoglobulin light chain. **Nature 239** : 117-120.

Misek D.E., Bard E. & Rodriguez-Boulan E. (1984). Biogenesis of epithelial cell polarity : intracellular sorting and vectorial exocytosis of an apical plasma membrane glycoprotein. **Cell 39** : 537-546.

Mize N.K., Andrews D.N. & Lingappa V.R. (1986). A stop transfer sequence recognizes receptors for nascent chain translocation across the endoplasmic reticulum membrane. **Cell 47** : 711-719.

Ness S.A. & Weiss R.L. (1987). Carboxyl terminal sequences influence the import of mitochondrial protein precursors "in vivo". **Proc. Natl. Acad. Sci. USA 84** : 6992-6996.

Pagano R.E. & Sleight R.G. (1985). Defining lipid transport pathways in animal cells. **Science 229** : 1051-1057.

Pages J.M. (1983). Biosynthèse et exportation des protéines de l'enveloppe d'Escherichia coli. **Biochimie 65** : 531-541.

Pain D., Kanwar Y.S. & Blobel G. (1988). Identification of a receptor for protein import into chloroplasts and its localization to envelope contact zones. **Nature 331** : 232-237.

Palade G.E. (1975). Intracellular aspects of the process of protein synthesis. **Science 189** : 347-358.

Palade G.E. (1983). Membrane biogenesis : An overview in "**Methods Enzymol.**", Vol. 96, pp XXIX-LV (S. Fleischer & B. Fleischer eds) Acad. Press.

Pfanner N., Muller H.K., Harmey M.A. & Neupert W. (1987). Mitochondrial protein import : involvement of the mature part of a cleavable precursor protein in the binding to receptor sites. **EMBO J. 6** : 3449-3454.

Pfanner N. & Neupert W. (1986). Transport of F_1-ATPase subunit β into mitochondria depends on both a membrane potential and nucleoside triphosphates. **FEBS lett. 209** : 152-156.

Rietveld J. & De Kruijff B. (1986). Phospholipids as possible instrument for translocation of nascent proteins across biological membranes. **Bioscience Reports 6** : 775-782.

Roise D. & Schatz G. (1988). Mitochondrial presequences. J. Biol. Chem. 263 : 4509-4511.

Rothman J.E. & Kornberg R.D. (1986). An unfolding story of protein translocation. Nature 322 : 209-210.

Roubin R., Tencé M., Mencia-Huerta J.M., Arnoux B., Ninio E. & Benveniste J. (1983). A chemically defined monokine : macrophage derived platelet activating factor (PAF acether) in "Lymphokines" Vol 8 pp 249-273 (E. Pick ed) Acad Press N.Y.

Sabatini D.D., Kreibich G., Morimoto T. & Adesnik M. (1982). Mechanisms for the incorporation of proteins in membranes and organelles. J. Cell Biol. 92 : 1-22.

Sakaguchi M., Mihara K. & Sato R. (1987). A short amino terminal segment of microsomal cytochrome P-450 functions both as a insertion signal and as a stop transfer sequence. EMBO J. 6 : 2425-2431.

Scheckman R. (1985). Protein localization and membrane traffic in yeast in "Ann. Rev. Cell Biol." Vol 1 pp 115-143 (G. Palade, B.M. Alberts & J.A. Spudich eds). Ann. Rev. Inc, Palo Alto CA.

Schmidt M.F.G. (1982). Acylation of proteins : an new type of modification of membrane glycoproteins. Trends Biochem. Sci 7 : 322-324.

Schmidt G.W. and Mishkind M.L. (1986). The transport of proteins into chloroplasts, in "Ann. Rev. Biochem", Vol. 55, pp 879-912 (Richardson, Boyer, Dawid, Meister eds). Ann. Rev. Inc, Palo Alto CA.

Schwaiger M., Herzog V. & Neupert W. (1987). Characterization of translocation contact sites involved in the import of mitochondrial proteins. J. Cell Biol. 105 : 235-246.

Scow R.O. & Blanchette-Mackie E.J. (1985). Why fatty acid flow in cell membranes ? Prog. Lipid Res. 24 : 197-241.

Semenza G. (1988). Biosynthesis and mode of insertion of a stalked intrinsic membrane protein of the small intestinal brush border (this book).

Singer S.J., Maher P.A. & Yaffe M.P. (1987). On the transfer
of integral proteins into membranes. **Proc. Natl. Acad. Sci.
USA 84** : 1960-1964.

Steinmen R.M., Mellman I.S., Muller W.A. & Cohn Z.A. (1983).
Endocytosis and the recycling of plasma membrane. **J. Cell
Biol. 96** : 1-27.

Tzagoloff A. & Myers A.M. (1986). Genetics of Mitochondrial
biogenesis, in "**Ann. Rev. Biochem.**" Vol. **55** pp 249-286
(Richardson, Boyer, Dawid, Meister eds) Ann. Rev. Inc.
Palo Alto CA.

Vale R.D. (1987). Intracellular transport using microtubule
based motors, in "**Ann. Rev. Cell Biol.**" Vol 3, pp 347-378
(G. Palade, B.M. Alberts, J.A. Spudich eds). Ann. Rev. Inc.
Palo Alto CA.

Vale R.D., Scholey J.M. & Sheetz M.P. (1986). Kinesin :
possible biological roles for a new microtubule motor.
Trends Biochem. Sci. 11 : 464-468.

Van Loon A.P.G.M., Brändli A.W. & Schatz G. (1986). The
presequences of two imported mitochondrial proteins contain
information for intracellular and intramitochondrial
sorting. **Cell 44** : 801-812.

Vaz W., Goodsaid-Zaldvondo F. & Jacobson K. (1984). Lateral
diffusion of lipids and proteins in bilayer membranes. **FEBS
lett. 174** : 199-207.

Von Figura K. & Hasilik A. (1986). Lysosomal enzymes and
their receptors in "**Ann. Rev. Biochem.**" Vol. **55**, p,p 167-194
(Richardson, Boyer, Dawid, Meister eds) Ann. Rev. Inc. Palo
Alto CA.

Walsh K.A., Ericson L.H., Parmelee D.C. & Titani K. (1981).
Advances in protein sequencing A review of up to date,
including automated sequence analysis, in "**Ann. Rev.
Biochem.**". Vol. **50**, pp 261-284 (Snell, Boyer, Meister &
Richardson eds). Ann. Rev. Inc. Palo Alto CA.

Walter P. & Blobel G. (1981). Translocation of protein across the endoplasmic reticulum III. Signal recognition protein causes signal dependant and site-specific arrest of chain elongation that is released by microsomal membranes. J. Cell. Biol. 91 : 557-561.

Walter P., Gilmore R. & Blobel G. (1984). Protein translocation across the endoplasmic reticulum. Cell 38 : 5-8.

Walter P. & Lingappa V.R. (1986). Mechanisms of protein translocation across the endoplasmic reticulum membrane in "Ann. Rev. Cell Biol." Vol 2, pp 499-516 (G. Palade, B.M. Alberts & J.A. Spudich eds). Ann. Rev. Inc. Palo Alto CA.

Warren G. (1985). Membrane traffic and organelle division. Trends Biochem. Sci. 10 : 439-443.

Weich H., Sagstetter M., Müller G. & Zimmermann R. (1987). The ATP requiring step in assembly of M13 procoat protein into microsomes is related to preservation of transport competence of the precursor protein. EMBO J. 6 : 1011-1016.

Wickner W. (1983). M13 coat protein as a model of membrane assembly. Trends. Biochem. Sci. 8 : 90-94.

Wickner W. & Lodish H.F. (1985). Multiple mechanisms of insertion of proteins into and across membranes. Science 230 : 400-407.

Williamson P., Antia R. & Schlegel R.A. (1987). Maintenance of membrane phospholipid asymetry ; lipid-cytoskeletal interactions or lipid pump ? FEBS lett. 219 : 316-320.

Wirtz K.W.A. (1974). Transfer of phospholipids between Membranes. Biochim. Biophys. Acta 334 : 95-117.

Yost C.S., Hedgepth J. & Lingappa V.R. (1983). A stop transfer sequence confers predictable transmembrane orientation to a previously secreted protein in cell free systems. Cell 34 : 759-766.

Zachowski A., Fellmann P., Hervé P. & Devaux P.F. (1987).
    Labeling of human erythrocyte membrane proteins by photoac-
    tivatable radioiodinated phosphatidylcholine and phosphati-
    dylserine. A search for the amino phospholipid translocase.
    FEBS Lett. 223 : 315-320.
Zimmermann R. & Meyer D.I. (1986). A year of new insights into
    how protein cross membranes. Trends Biochem. Sci. 11 :
    512-515.

FLUID-PHASE ENDOCYTOSIS AND LYSOSOMAL ENZYME EXCRETION
TWO FACETS OF INTRACELLULAR MEMBRANE TRAFFIC

G. KLEIN, M. BOF, C. GONZALEZ and M. SATRE

1. Introduction

Endocytosis is the cellular phenomenon by means of which all eukaryotes, at least to some degree, internalize material in vesicles derived from invaginations of their own plasma membrane. In fluid-phase pinocytosis, the internalized fluid components do not bind to the plasma membrane, and entry is non-saturable with respect to concentration of compounds contained in the fluid. This is in contrast to receptor-mediated endocytosis where internalized ligands bind to specific cell surface receptors (for reviews see Besterman and Low, 1983 ; Steinman et al., 1983 ; Wileman et al., 1985). Endocytosis is associated with a very active membrane traffic between plasma membrane, endosomes and lysosomes, and detailed labeling and morphometric studies in amoebae and in other mammalian cells have directly demonstrated that endocytosis is an homeostatic phenomenon and that the relative amounts of membrane in the various endocytic compartments remain stable (Thilo, 1985). We fill focus here on some selected aspects of fluid-phase pinocytosis and on the associated secretion of lysosomal enzymes.

2. Fluide-phase pinocytosis

The general pathways of fluid-phase pinocytosis are depicted in the simple scheme shown in fig. 1 (Besterman and Low, 1983 ; Steinman et al., 1983 ; Duncan and Pratten, 1985 ; Thilo, 1985). Incoming coated- or uncoated vesicles (compartment 1) constitute the first compartment for fluid entry and they rapidly fuse with each other, thus decreasing their surface to volume ratio. The next participating compartment is constituted by the endosomes (compartment 2), which appear as size-heterogeneous prelysosomal organelles. Up to this stage, all the compartments are topologically equivalent to the

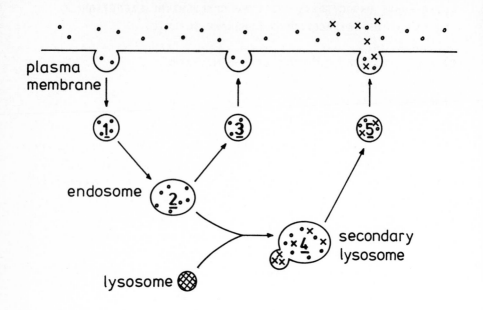

Figure 1 : Scheme of vesicular traffic during fluid-phase pinocytosis.

Two stages of membrane recycling are indicated : (3) during fusion of pinocytic vesicles and processing endosomes ; (5) recycling via secondary lysosome membrane pool.

(o) fluid-phase marker molecule ; (X) lysosomal hydrolytic enzymes.

plasma membrane and they contain no soluble lysosomal enzymes, except for any enzymes that could be re-internalized following secretion (see later). It has not been demonstrated conclusively whether endosomes are directly derived by successive fusion steps from the internalized plasma membrane or if they constitute pre-existing vesicular docking organelles. Fusion of the primary lysosomes with the endosomes to form secondary lysosomes (compartment 4) allows the hydrolytic lysosomal enzymes to come into contact with internalyzed material. The membrane of secondary lysosomes contains permeases originating from lysosomal membrane (Reijngoud and Tager J.M., 1977 ; Hales et al., 1984). In cells such as the amoeba Dictyostelium, which contain a limited complement of permeases or even no permease for sugars or amino acids in their plasma membrane, it is only at this stage that the endocytosed nutrients can cross the membrane into the cytosolic space. Membrane retrieval for recycling to the plasma membrane occurs both before any fusion with primary lysosomes (compartment 3) and after the formation of secondary lysosomes (compartment 5).

FITC-dextran as a fluid-phase marker

Pinocytosis is measured by following the entry into the cells of a fluid-phase marker. Several criteria have been proposed for the choice of a satisfactory fluid-phase marker, such as the obvious properties of non-toxicity, intra- and extracellular stability, absence of hydrophobic or ionic adsorption to the cell constituents, but the final choice in a given cellular system always remains guided by practical experience (for a recent discussion, see Wiley and McKinley, 1987). The use of fluorescein-labeled high molecular weight dextran (FITC-dextran, see fig. 2) with a low degree of substitution by fluorescein has been validated in many different cell types. FITC-dextran is easily detectable with a great sensitivity by fluorescence spectroscopy. In addition,

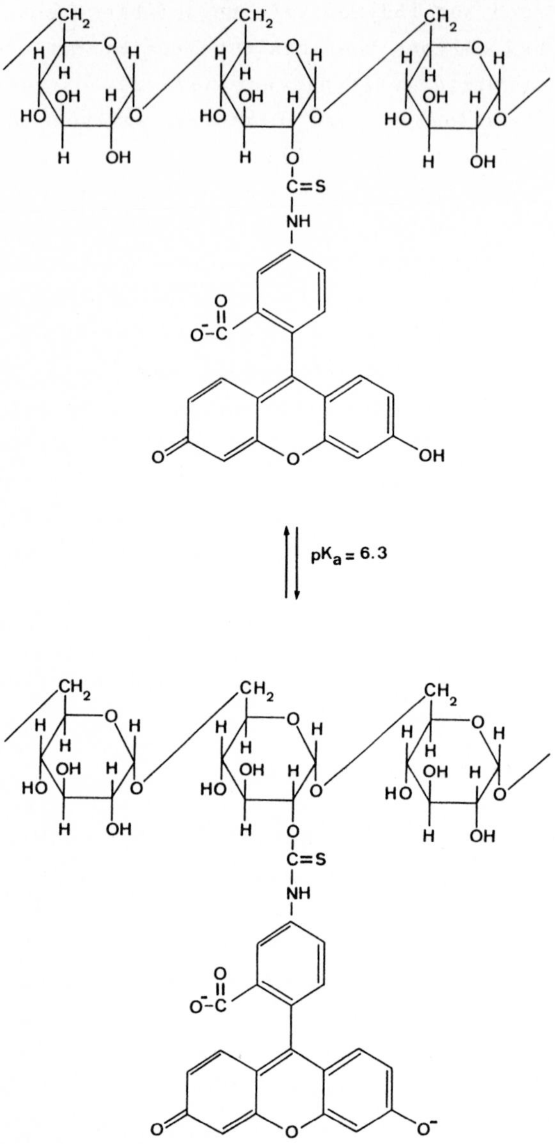

Figure 2 : Acid-base equilibrium of FITC-dextran
In the physiological range, the apparent pKa of the equilibrium of fluorescein is about 6.3.

the marked pH dependence of its fluorescence excitation spectrum can be used to measure the pH of the endosomal compartment (Ohkuma and Poole, 1978 ; Geisow, 1984).

The possible contamination of commercial batches of FITC-dextran by low molecular weight reactive fluorescent compounds should always be carefully checked (Preston et al., 1987) and is especially critical when measurements of low rates of pinocytosis are contemplated. In case of doubts about the purity of FITC-dextran, synthesis should be considered and an easy protocol has been described (De Belder and Granath, 1973). This problem of purity of the fluid-phase marker is not restricted to FITC-dextran and has also been described for several other compounds, including (^{14}C)-sucrose, another classical marker (Besterman et al., 1981 ; Wiley and McKinley, 1987).

Kinetics of fluid-phase pinocytosis

Rates and extent of fluid-phase pinocytosis are cell-type dependent. In highly endocytic cells, such as the amoebae _Entamoeba histolytica_ and _Dictyostelium discoideum_, where endocytosis is basically a feeding process, a pinocytosis plateau is rapidly reached after a linear phase of uptake. The plateau corresponds to a dynamic equilibrium between entry and exit of fluid from the cell, and not to an arrest of pinocytosis. Kinetically, these amoebae appear to have a unique endosomal compartment (Aley et al., 1984 ; Klein and Satre, 1986).

In higher eukaryotic cells, such as fibroblasts or hepatocytes, where fluid-phase pinocytosis rates are slower, modeling of the process led to the identification of two kinetically distinguishable sequential compartments : the prelysosomal compartment of small size which is rapidly filled and followed by the secondary lysosomes, a compartment of larger size with a much lower filling rate (Besterman et al.,

1981 ; Thilo, 1985 ; Scharschmidt et al., 1986). The kinetic differences between the various cell types can be interpreted within the framework of the same model (see fig. 1) by considering different residence times and transfer rates between the successive compartments.

Efficiency of fluid-phase pinocytosis is greatly increased by a concentration phenomenon of the internalyzed medium (Swanson et al., 1986 ; Klein and Satre, 1986).

Endosomal pH

In many cells, evidence has accumulated showing that the interior of all the endocytic compartments is maintained at an acidic pH. Several membrane-bound H^+- translocating ATPases appear to have a central function in pH regulation all along the endosomal pathway (for reviews see Maxfield, 1985 ; Mellman et al., 1986 ; Anderson and Orci, 1988).

Fluorescein-labeled fluid-phase markers are the molecules of choice for pH determination in endosomal compartments of isolated cells (Ohkuma and Poole, 1978 ; Geisow, 1984). For accurate pH measurements and to avoid potential artifacts, calibration curves are best conducted in conditions as close as possible to the in vivo situation. To this effect, after loading with FITC-dextran and removal of extracellular fluid-phase marker, all intracellular compartments are fully equilibrated to various extracellular pH values, either byusing pH buffers containing both weak acid and base, for example, by addition of ammonium acetate, or alternatively by using pH buffers containing a high K^+ concentration together with nigericin, a K^+ - H^+ ionophore, to dissipate any transmembrane pH gradients (Yamashiro and Maxfield, 1987).

In most cells, the endosomal compartment is maintained at an average pH of about 5.5-6, somewhat less acidic than primary lysosomes (pH < 5). The pH gradient between the cytosol itself and endosomal interior space is thus close to

1.5 to 2 pH units. The importance of such pH gradients for proper functioning of the endocytic machinery has been emphasized (Maxfield, 1985 ; Mellman et al., 1986 ; Anderson and Orci, 1988). An interesting exception is found in the pathogenic amoeba Entamoeba histolytica, where it has been shown that the endosomal compartment is non-acidified and is in simple equilibrium with external pH (Aley et al., 1984).

3. Exocytosis of lysosomal enzymes

Once thought to be a phenomenon of rare occurrence, most eukaryotic cells are now known to secrete constitutively lysosomal enzymes. The exact biological significance of this phenomenon, occurring in the absence of cell lysis and highly selective for lysosomal enzymes, is still unexplained. This energy-requiring process has been observed and studied in some free-living eukaryotic micro-organisms where it is very active, like the amoeba Acanthamoeba castellanii (Hohman and Bowers, 1984, 1985), the slime mould Dictyostelium discoideum (Dimond et al., 1981), the two ciliate protozoa Tetrahymena pyriformis and T. thermophila (Banno et al., 1987 ; Hünseler et al., 1987) and is also present in several mammalian cell types (Leoni et al., 1985 ; Hermelin et al., 1988 ; Schnyder & Baggiolini, 1980).

Several different lysosomal enzyme secretion mechanisms have been demonstrated. In some well-characterized cases, secretion occurs as a result of errors in intracellular targeting, and the lysosomal proteins without their correct phosphomannosyl sorting signals are not delivered into primary lysosomes and instead become secreted (Creek and Sly, 1984 ; Von Figura and Hasilik, 1986). Secretion also arises by another mechanism which will be considered here and is a direct consequence of normal membrane recycling during endocytosis. Protein biosynthesis is not a prerequisite for this

secretion pathway, as the secreted lysosomal enzymes are
simply those that are passively trapped in shuttle vesicles
that cycle back membrane from secondary lysosomes to the cell
plasma membrane (sequence 4 - - > 5, fig. 1). As such, the
study of lysosomal enzymes secretion may provide interesting
clues to the details of endocytic machinery.

Secretion heterogeneity of lysosomal hydrolases

Different lysosomal enzymes are secreted with unequal
kinetics and on this basis have been classified in two or even
three groups having different rates and extent of secretion.
Among all the lysosomal enzymes examined, no general pattern
seems to emerge, although in three micro-organisms studied :
Acanthamoeba, Dictyostelium and Tetrahymena, the hexosamini-
dase always belongs to the group of rapidly excreted lysosomal
enzymes, in contrast to acid phosphatase which is always
representative of the group showing a slower secretion (Hohman
and Bowers, 1984 ; Dimond et al., 1981 ; Banno et al., 1987).

The heterogeneity in lysosomal hydrolases secretion is
not easily explained on the basis of current data. Neither
supposing the mere existence of structurally and functionally
distinct primary lysosomes, nor the postulated different
binding affinities of various lysosomal enzymes for endosomal
membranes leading to differential release and recapture, can
fully account for secretion heterogeneity. Several secretory
mutants have been isolated in Dictyostelium and in Tetrahymena
(Dimond et al., 1983 ; Hünseler et al., 1987). Their
characterization is still preliminary and although they have
not yet revealed decisive clues about the mechanisms of
lysosomal enzyme secretion these genetic studies are clearly
of great promise.

Effect of sugars on lysosomal enzyme secretion

Another experimental approach to investigating the lysosomal excretion pathways is to use perturbing conditions. Secretion of lysosomal enzymes is drastically stimulated by non-permeant and non-metabolizable sugars. Sucrose is often chosen as a model compound to overload the endosomal system (Fell and Dingle, 1966 ; Crean and Rossomando, 1979 ; Dimond et al., 1981 ; Muir and Bowyer, 1984 ; Sheshadri et al., 1986).

In Dictyostelium amoeba, where detailed studies have been conducted, more than 90 % of the total cellular activity of several lysosomal enzymes, including glycosidases and acid phosphatase, becomes extracellular within a few hours in the presence of sucrose. The kinetics of excretion are complex, with an initial lag period followed by a phase of rapid secretion, and a final plateau with little excretion when the pool of intracellular lysosomal enzymes has almost reached exhaustion.

Sucrose is internalized by fluid-phase pinocytosis and reaches the secondary lysosome. There, it cannot cross the membrane to gain access to the cytosol and it remains osmotically active leading to valuolated cells (Cohn and Ehrenreich, 1969 ; Swanson et al., 1986). At the same time, mobilization of primary lysosomes and fusion with endosomes are strongly enhanced. Vacuolation and lysosomal enzymes secretion are not necessarily linked events. Cellular vacuolation by weak bases does not led to secretion and a Tetrahymena mutant constitutively blocked in the secretion of lysosomal enzymes has a highly vacuolated phenotype (Okhuma and Poole, 1981 ; Hünseler et al., 1987).

4. Conclusion

Fluorescent pH-sensitive fluid-phase markers such as FITC-dextran are important reagents for measuring pinocytosis and for improving our understanding of the functioning and regulation of pinocytic fluid flow through eukaryotic cells, and the functions of different endosomal compartments. Lysosomal enzyme excretion in relation to membrane recycling during pinocytosis is an efficient process of particular interest, and represents the mirror image of the pathway of incoming pinocytic vesicles.

References

Aley S.B., Cohn Z.A. and Scott W.A. (1984). Endocytosis in Entomoeba histolytica. Evidence for a unique non-acidified compartment. J. Exp. Med. 160 : 724-737.

Anderson R.G.W. and Orci L. (1988). A view of acidic intracellular compartments. J. Cell Biol. 106 : 539-543.

Banno Y., Sasaki N. and Nozawa Y. (1987). Secretion heterogeneity of lysosomal enzymes in Tetrahymena pyriformis. Exp. Cell Res. 170 : 259-268.

Besterman J.M., Airhart J.A., Woodworth R.C. and Low R.B. (1981). Exocytosis of pinocytosed fluid in cultured cells : kinetic evidence for rapid turnover and compartmentation. J. Cell Biol. 91 : 716-727.

Besterman J.M. and Low R.B. (1983). Endocytosis : a review of mechanisms and plasma membrane dynamics. Biochem. J. 210 : 1-13.

Cohn Z.A. and Ehrenreich B.A. (1969). The uptake, storage and intracellular hydrolysis of carbohydrates by macrophages. J. Exp. Med. 129 : 201-225.

Crean E.V. and Rossomando E.F. (1979). Effects of sugars on glycosidase secretion in Dictyostelium discoideum. J. Gen. Microbiol. 110 : 315-322.

Creek K.E. and Sly W.S. (1984). The role of the phosphomannosyl receptor in the transport of acid hydrolases to lysosomes. in "**Lysosomes in Biology and Pathology**", (eds.
Dingle J.T. and Dean R.T.) vol. **7**, pp. 63-88. Elsevier
Science Publishers, Amsterdam, Holland.

De Belder A.N. and Granath K. (1973). Preparation and properties of fluorescein-labelled dextrans. **Carbohydrate Res.**
30 : 375-378.

Dimond R.L., Burns R.A. and Jordan K.B. (1981). Secretion of
lysosomal enzymes in the cellular slime mold, Dictyostelium
discoideum. **J. Biol. Chem. 256** : 6565-6572.

Dimond R.L., Knecht D.A., Jordan K.B, Burns R.A. and Livi G.P.
(1983). Secretory mutants in the cellular slime mold
Dictyostelium discoideum. **Methods Enzymol. 96** : 815-828.

Duncan R. and Pratten M.K. (1985). Pinocytosis : mechanism and
regulation. in "**Mononuclear Phagocytes : Physiology and
Pathology**", (eds. Dean R.T. and Jessup W.) pp. 27-51.
Elsevier Science Publishers, Amsterdam, Holland. .

Fell H.B. and Dingle J.T. (1966). Extracellular release of
lysosomal enzymes in response to sucrose. **Biochem. J. 98** :
40P.

Geisow M.J. (1984). Fluorescein conjugates as indicators of
intracellular pH. A critical evaluation. **Exp. Cell Res.**
150 : 29-35.

Hales C.N., Docherty K. and Maguire G.A. (1984). Sugar
transport in lysosomes. in "**Lysosomes in Biology and
Pathology**", (eds. Dingle J.T. and Dean R.T.). **Vol. 7**, pp.
165-174. Elsevier Science Publishers, Amsterdam, Holland.

Hermelin B., Cherqui G., Bertrand F., Wicek D., Paul A.,
Garcia I. and Picard J. (1988). Phorbol ester-induced
protein kinase C translocation and lysosomal enzyme
release in normal and cystic fibrosis fibroblasts. **FEBS-
Lett. 229** : 161-166.

Hohman T.C. and Bowers B. (1984). Hydrolase secretion is a consequence of membrane recycling. J. Cell Biol. 98 : 246-252.

Hohman T.C. and Bowers B. (1985). Vacuolar pH is one factor that regulates hydrolases secretion. Eur. J. Cell Biol. 39 : 475-480.

Hünseler P., Scheidgen-Kleyboldt G. and Tiedke A. (1987). Isolation and characterization of a mutant of Tetrahymena thermophila blocked in secretion of lysosomal enzymes. J. Cell Sci. 88 : 47-55.

Klein G. and Satre M. (1986). Kinetics of fluid-phase pinocytosis in Dictyostelium discoideum amoebae. Biochem. Biophys. Res. Commun. 138 : 1146-1152.

Leoni P., Dean R.T. and Jessup W. (1985). Secretion of hydrolases by mononuclear phagocytes. In "Mononuclear Phagocytes : Physiology and Pathology", (eds. Dean R.T. and Jessup W.), pp. 181-202. Elsevier Science Publishers, Amsterdam, Holland.

Maxfield F.R. (1985). Acidification of endocytic vesicles and lysosomes. in "Endocytosis" (eds. Pastan I. and Willingham M.C.). pp. 235-257. Plenum Press, New-York.

Mellman I., Fuchs R. and Helenius A. (1986). Acidification of the endocytic and exocytic pathways. Ann. Rev. Biochem. 55 : 663-700.

Muir E.M. and Bowyer D.E. (1984). Inhibition of pinocytosis and induction of release of lysosomal contents by lysosomal overload of arterial smooth muscle cells in vitro. Atherosclerosis 50 : 85-92.

Ohkuma S. and Poole B. (1978). Fluorescence probe measurements of the intralysosomal pH in living cells and the perturbation of pH by various agents. Proc. Natl. Acad. Sci. USA 75 : 3327-3331.

Ohkuma S. and Poole B. (1981). Cytoplasmic vacuolation of mouse peritoneal macrophages and the uptake into lysosomes of weakly basic substances. J. Cell Biol. 90 : 656-664.

Preston R.A., Murphy R.F. and Jones E.W. (1987). Apparent
    endocytosis of fluorescein isothio-cyanate-conjugated
    dextran by Saccharomyces cerevisiae reflects uptake of low
    molecular weight impurities, not dextran. J. Cell Biol.
    105 : 1981-1987.
Reijngoud D.-J. and Tager J.M. (1977). The permeability
    properties of the lysosomal membrane. Biochim. Biophys.
    Acta 472 : 419-449.
Scharschmidt B.F., Lake J.R., Renner E.L., Licko V. and Van
    Dyke R.W. (1986). Fluid-phase pinocytosis by cultured rat
    hepatocytes and perfused rat liver : implications for
    plasma membrane turnover and vesicular trafficking of
    fluid-phase markers. Proc. Natl. Acad. Sci. USA 83 :
    9488-9492.
Schnyder J. & Baggiolini M. (1980). Secretion of lysosomal
    enzymes by macrophages in "Mono-nuclear phagocytes :
    functional aspects, part 2", (ed. Van Furth R.), pp.
    1369-1384. Martinus Nijhoff publishers The Hague Holland.
Sheshadri J., Cotter D.A. and Dimond R.L. (1986). The
    characterization and secretion pattern of the lysosomal
    trehalases of Dictyostelium discoideum. Exp. Mycol. 10 :
    131-143.
Steinman R.M., Mellman I.S., Muller W.A. and Cohn Z.A. (1983).
    Endocytosis and the recycling of plasma membrane. J. Cell
    Biol. 96 : 1-27.
Swanson J., Yrinec B., Burke E., Bushnell A. and Silverstein
    S.C. (1986). Effect of alterations in the size of the
    vacuolar compartment on pinocytosis in J774.2 macrophages.
    J. Cell Physiol. 128 : 195-201.
Thilo L. (1985). Quantification of endocytosis-derived
    membrane traffic. Biochim. Biophys. Acta 822 : 243-266.
Von Figura K. and Hasilik A. (1986). Lysosomal enzymes and
    their receptors. Ann. Rev. Biochem. 55 : 167-193.

Wileman T., Harding C. and Stahl P. (1985). Receptor-mediated endocytosis. **Biochem. J.** *232* : 1-14.

Wiley H.S. and McKinley D.N. (1987). Assay of growth factor stimulation of fluid-phase endocytosis. **Methods Enzymol.** **146** : 402-417.

Yamashiro D.J. and Maxfield F.R. (1987). Kinetics of endosome acidification in mutant and wild-type Chinese hamster ovary cells. **J. Cell Biol.** **105** : 2713-2721.

THE NEUTROPHIL LEUKOCYTE. PROPERTIES AND MECHANISM OF ACTIVATION

M. BAGGIOLINI, D.A. DERANLEAU, B. DEWALD, M. THELEN, V. VON TSCHARNER, M.P. WYMANN

The neutrophil leukocytes constitute the largest population of white blood cells. Their main function is to defend the host organism from microbial invasion. The properties required for this function are chemotactic responsiveness, mobility and the ability to phagocytose and to release microbicidal products. Microorganisms which colonize a tissue are sensed by the neutrophils through chemotactic molecules generated upon infection. In response to such stimuli, neutrophils adhere to the endothelia of the microvessels of the infected area, leave the blood and migrate toward the invaders, eventually phagocytosing and killing them. Killing requires the activation of NADPH-oxidase, a plasma membrane enzyme that generates superoxide (Babior, 1978 ; Baggiolini, 1984), and is aided by the release of enzymes and other storage proteins from the azurophil and specific granules (Baggiolini & Dewald, 1984)). The event initiating this host defense process is the activation of the neutrophils via chemotactic agonists and phagocytosis receptors. The mechanism of activation is generally studied using chemotactic agonists rather than phagocytosable particles. As soluble molecules, the agonists have the advantage of acting instantaneously and uniformly on all target cells. Four such agonists, each acting via a selective receptor, have been characterized in recent years. They are the anaphylatoxin C5a formed upon complement activation via the classical or the alternative pathway (Fernandez et al., 1978), N-formyl-Met-Leu-Phe (fMLP) and other N-formylmethionyl peptides of bacterial origin (Showell et al., 1976), and two bioactive lipids, platelet-activating factor (PAF) and Leukotriene B_4 (LTB_4) (Ingraham et al., 1982 ; Ford-Hutchinson et al., 1980 ; Baggiolini et al., 1988). PAF and LTB_4 are produced by activated phagocytes and can therefore function as amplifiers

of the cellular response (Dewald & Baggiolini, 1985). A novel
chemotactic peptide produced by stimulated human monocytes
was recently described, and was termed NAF for neutrophil-activating factor or MDNCF for monocyte-derived neutrophil chemotactic factor (Walz et al., 1987 ; Yoshimura et al., 1987 ;
Peveri et al., 1988). All of the known chemotactic agonists
induce shape changes, exocytosis and the respiratory burst,
functional responses that depend on a rise of cytosolic free
calcium ($[Ca^{2+}]_i$) and that can be followed in real time with
appropriate instrumentation.

The mechanism of neutrophil activation, in particular
the process of transduction of agonist signals, is largely
unknown. We have recently studied two aspects of this
process, the stimulus-induced changes in $[Ca^{2+}]_i$ and the
initiation of the respiratory burst.

Agonist-induced $[Ca^{2+}]_i$ changes

A broad-band filter fluorimeter with a response time
of 30 msec, was used for the real-time recording of $[Ca^{2+}]_i$
changes induced by stimulation of human neutrophils (Von
Tscharner et al., 1986). The cells were loaded with quin-2 or
fura-2, and were resuspended shortly before stimulation in an
isotonic saline buffer containing 1 mM $CaCl_2$ or 1 mM EGTA and
no $CaCl_2$. They were then stimulated at 37° with fMLP, PAF or
the calcium ionophore ionomycin in the stirred cuvette of the
fluorimeter. Full mixing of the stimulus in the cell suspension was obtained within 0.3 sec. All three stimuli induced a
rise in $[Ca^{2+}]_i$, both in the absence and presence of calcium
in the extracellular medium. In the latter case, $[Ca^{2+}]_i$ rose
at a faster rate and to higher levels, confirming that
calcium influx through the plasma membrane contributed to the
change. Two kinetic features of the $[Ca^{2+}]_i$ response, however, were independent of extracellular calcium, the onset

time, i.e. the time elapsing between stimulation and the
beginning of the fluorescence increase, and the apparent
duration of the response, as expressed by the time required
to reach 90 % of the maximum fluorescence level. The high
time resolution of the instrument revealed a lag of about 1
to 3 sec preceding the $[Ca^{2+}]_i$ rise induced by the two
receptor agonists. By contrast, following addition of ionomycin $[Ca^{2+}]_i$ rose immediately, i.e. within less than 0.3 sec.
The duration of the lag increased when the concentration of
fMLP or PAF was progressively decreased from 100 to 1 nM (Von
Tscharner et al., 1986). These results suggested the involv-
ment of a rate-limiting process that controls agonist-induced
calcium release from internal stores and calcium influx
across the plasma membrane.

Patch-clamp experiments (Hamill et al., 1981) were
performed to study the mechanism of calcium influx through
the neutrophil plasma membrane (Von Tscharner et al., 1986).
In cell-attached and in isolated, inside-out patches, different current levels revealed two types of channels passing
Ca^{2+}. Both channels appear to qualify as cation channels
rather than as selective calcium channels since Na^+ and K^+
were passed as well. Virtually no channel openings were
recorded in resting cells or upon stimulation with fMLP
through the recording pipette. The latter observation suggests that receptor-operated channels are not involved in the
activation of neutrophils by chemotactic agonists. Channel
openings were frequent (i) in cell-attached patches following
stimulation of the cells by adding fMLP to the bath, and (ii)
in isolated inside-out patches or saponin-permeabilized cells
exposed to calcium concentrations (in the bath) above 0.1 µM.
No channel activity was observed in fMLP-stimulated
calcium-depleted cells. Using excised inside-out patches, the
possibility that inositol 1, 4, 5-triphosphate (IP_3) could
open calcium channels in the plasma membrane was investigated. Although such a mechanism is believed to lead to the

release of calcium from intracellular stores (Prentki et al., 1984), no electrical activity could be detected unless the calcium concentration in the bath was increased above resting $[Ca^{2+}]_i$ levels.

A possible mechanism for the $[Ca^{2+}]_i$ rise

It is generally believed that IP_3 acts as a second messenger in neutrophils as it does in many other types of cells (Berridge, 1984). In fact, IP_3 induces calcium release in permeabilized neutrophils (Prentki et al., 1984) and **B. pertussis** toxin, which is known to block receptor-mediated activation of neutrophils (Molski et al., 1984), prevents both the formation of IP_3 and a $[Ca^{2+}]_i$ rise in agonist-stimulated cells (Ohta et al., 1985 ; Okajima et al., 1985). Our results concur with this notion and suggest the following mechanism for the agonist-induced rise in $[Ca^{2+}]_i$. Upon binding with an agonist, the receeport interacts with a GTP-binding protein which activated the phosphatidylinositol specific phospholipase C (Cockcroft and Gomperts, 1985 ; Smith et al., 1985) and leads to the liberation of IP_3. While the action of IP_3 on intracellular calcium stores is well documented (Spaet et al., 1986), our patch-clamp experiments suggest that IP_3 does not affect calcium influx across the plasma membrane. Thus, if indeed agonist-induced $[Ca^{2+}]_i$ changes are mediated by IP_3, the initiating event must be at the level of the internal calcium stores. Calcium release from these stores enhances $[Ca^{2+}]_i$, which causes the opening of plasma membrane cation channels and the influx of calcium. It is not unlikely that the internal calcium stores are located in the ectoplasmic region - close to the source of the second messenger IP_3 and close to plasma membrane channels that were shown by our patch-clamp experiments to be sensitive to a $[Ca^{2+}]_i$ rise (Krause and Lew, 1987).

The lag preceding the $[Ca^{2+}]_i$ rise, which was demonstrated in our experiments, is a characteristic feature of the
response of neutrophils to receptor agonists. No lag was
observed upon challenge with ionomycin, indicating that
calcium mobilization in itself (release from internal stores
and/or influx from the external medium) is not time-limiting.
Rather, the lag could reflect the time required for the
generation of threshold levels of IP_3. The lag was dependent
on the agonist concentration and therefore on receptor
occupancy, which most likely controls the extent of phospholipase C activation. The transient nature of agonist-induced
$[Ca^{2+}]_i$ changes implies a rapid cessation of calcium supply,
which most probably consequence of the metabolic transformation of IP_3 (Berridge, 1984), and the reuptake and/or efflux
of calcium reestablishing resting $[Ca^{2+}]_i$ levels (Von Tscharner et al., 1986).

The respiratory burst

Two elements are believed to be essential for the
activation of the NADPH-oxidase in intact neutrophils, enhanced $[Ca^{2+}]_i$ and active protein kinase C. A role for protein
kinase C is suggested by the fact that phorbol esters, e.g.
phorbol myristate acetate (PMA), and permeant diacylglycerols
elicit the respiratory burst (Repine et al., 1974 ; Kaibuchi
et al., 1983). The requirement for calcium is suggested by
the fact that in calcium-depleted neutrophils (where $[Ca^{2+}]_i$
cannot be enhanced) receptor agonists fail to induce a
respiratory burst (Grzeskowiak et al., 1986). It is likely
that upon stimulation with à receptor agonist, enhanced
$[Ca^{2+}]_i$ is needed, together with diacylglycerol formation, to
turn on protein kinase C. The transduction sequence that can
be assumed on the basis of the information gathered in many
laboratories is illustrated schematically in Fig. 1. Active
NADPH-oxidase requires ongoing agonist receptor binding

(Sklar et al., 1985), the interaction between the receptor and a GTP-binding protein(Okajima et al., 1985) and the activation of phosphatidylinositol-specific phospholipase C (Smith et al., 1985). At this level of the sequence, products are delivered which bring about the translocation of protein kinase C to the plasma membrane and its subsequent activation (Wolf et al., 1985 ; Horn and Karnovsky, 1986). The plausibility of the sequence illustrated is supported by results obtained under conditions indicated in the scheme. Signal transduction is inhibited or prevented by receptor antagonists or pretreatment of the cells with **B. pertussis** toxin, showing that receptor occupancy and functional GTP-binding proteins are essential. As already indicated, calcium-depleted cells do not transduce agonist signals leading to the activation of NADPH-oxidase, indicating that a $[Ca^{2+}]_i$ rise is necessary. The role of IP_3 in the latter event is suggested by the ability of exogenous IP_3 to induce the liberation of calcium in permeabilized neutrophils. Finally the role of protein kinase C is suggested by the fact that the transduction sequence can be short-cutted by protein kinase C ligands like PMA.

Respiratory burst induction by chemotactic agonists, PMA and ionomycin

The mechanism of NADPH-oxidase activation has been studied by comparing the onset time of the respiratory burst induced by different types of stimuli (Wymann et al., 1987). The time elpasing between stimulation and the first appearance of oxidase products, i.e. superoxide or H_2O_2, is a reflection of the time needed for signal transduction and can therefore be a useful value in the analysis of the mechanisms of neutrophil activation. Such a study demands a very sensitive assay for the response to be measured. We opted. for a chemiluminescence assay of H_2O_2 which arises from the instantaneous dismutation of superoxide. Under the conditions

of our assay (Wymann et al., 1987), stimulus-induced chemiluminescence was directly proportional to the rate of H_2O_2
generation and thus to the NADPH-oxidase activity.

The chemiluminescence response curves obtained with
different agonists varied considerably. Stimulation with C5a
yielded about 1/3 of the H_2O_2 produced upon stimulation with
fMLP, while PAF and LTB_4 yielded only as little as 1/20 to
1/50 of that amount (Wymann et al., 1987). Despite these
differences, all agonist-induced chemiluminescence responses
showed similar kinetic features. The onset time of the
responses was identical, and the slope of the chemiluminescence progress curves at the first inflection point
extrapolated back virtually to the same point in time
(Table 1 in Wymann et al., 1987). Maximum effective agonist
concentrations were used for these experiments, but similar
results were obtained at lower concentrations, as shown for
fMLP (Wymann et al., 1987).

The values presented in Table 1 characterize the
initiation of the respiratory burst response : the onset time
is a measure of the time required for signal transduction,
and the time-axis intercept of the maximum positive slope of
the H_2O_2 production curve represents the reciprocal of the
apparent first-order rate constant for the generation of
active NADPH-oxidase. The similarities in the onset time of
the response and in the rate of oxidase activation strongly
suggest that signals perceived by the neutrophil through
different receptors are transduced by a similar or possibly
even identical process. Once initiated, the response appears
to proceed at its own intrinsic pace, i.e. independently of
the type and the concentration of the triggering agonist. The
major differences in the overall respiratory burst responses
indicate that, in contrast to its activation, deactivation of
the NADPH-oxidase depends markedly on the type of agonist
used.

The chemiluminescence responses induced by ionomycin or PMA were quite different. The onset time was at least 3-5 times longer than that observed on stimulation with the agonists, ant the apparent rate constants for the generation of NADPH-oxidase activity were much smaller (0.05 to 0.07 sec^{-1} or about 1/4th of the values obtained with agonists).

Responses to combined stimuli

The differences in onset time and apparent rate constant observed indicate the existence of separate transduction mechanisms for the induction of the respiratory burst with agonists or PMA and suggest the possibility of positive or negative interferences upon use of these stimuli in combination. A synergism between protein kinase C activators and chemotactic agonists was shown by previous work from our laboratory (Dewald et al., 1984). Superoxide production in response to fMLP was found to be faster and more extensive when the cells were pretreated with 1-oleoyl-2-acetyl glycerol (OAG). Using chemiluminescence, we have been able to quantitate the effect of combined stimulation on the onset of the respiratory burst. The neutrophils were first treated with low concentrations of PMA or ionomycin and then, before a response became detectable, with fMLP. The pretreatment shortened the onset time of H_2O_2 production induced by fMLP to about half, i.e. from an average of 2.5 to 0.7-1.2 sec. Similar results were obtained with C5a, PAF or LTB_4 instead of fMLP, and OAG instead of PMA. A significant shortening of the onset time was also observed when the cells were stimulated in sequence with PMA and ionomycin. The onset time of the response to the latter combination, however, was still markedly longer than that observed upon stimulation with receptor agonists alone, indicating that the signal transduction process initiated by an agonist cannot be mimiked by concomitant protein kinase C activation and calcium influx.

In the above experiments, the second stimulus was always added before the appearance of a response to the first one. It was possible, however, to elicit a respiratory burst with a chemotactic agonist in neutrophils that were already producing H_2O_2 following stimulation with PMA or OAG. Under these conditions, the onset time of the agonist response fell below the time resolution of our detection system (Wymann et al., 1987). Using this protocol, we compared the onset of agonist-dependent $[Ca^{2+}]_i$ rise and H_2O_2 production. Neutrophils were loaded with fura-2, and were stimulated first with a low concentration of PMA (eliciting low levels of H_2O_2 formation) and then with fMLP. Simultaneous chemiluminescence and fura-2 fluorescence recordings showed that the onset of the respiratory burst was immediate while the $[Ca^{2+}]_i$ rise was delayed by about a sec, indicating that activation of NADPH-oxidase by the receptor agonist occurred without appreciable changes in $[Ca^{2+}]_i$. A time dissociation of the two signals was also observed when neutrophils responding to PMA were stimulated with ionomycin : the rise in $[Ca^{2+}]_i$ was immediate, in agreement with previous results (Von Tscharner et al., 1986), but H_2O_2 production was delayed by 6-7 sec.

Conclusions

Our results suggest that two transduction sequences are necessary for the induction of the respiratory burst by receptor agonists. One sequence is calcium- and protein kinase C-dependent and rate limiting, while the other is comparatively fast and appears calcium-independent. Both sequences are sensitive to **B. pertussis** toxin and are probably branching off downstream of G-proteins. A scheme of this transduction system is shown in Fig. 2. Agonists acting on different receptors engage a common type of G-proteins. Activation of phospholipase C delivers diacylglycerol and leads (via IP_3) to a rise in $[Ca^{2+}]_i$ and to the activation of protein kinase C. This is not possible in calcium-depleted cells which do not respond to

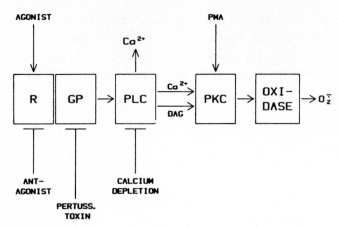

Figure 1 : Calcium- and protein kinase C-dependent presumed signal transduction sequence in human neutrophils. Activating and inhibiting measures are indicated. Outward arrow indicates step leading to release of cytosolic free calcium.

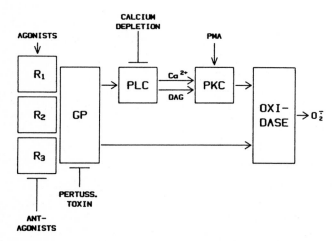

Figure 2 : Presumed branching transduction sequence of agonist-dependent signals in human neutrophils.

agonists unless PMA or other protein kinase C ligands are supplied. Two observations suggest that both sequences must be functional and must operate in concert for transducing receptor-dependent signals : (i) Calcium depletion, which only affects the upper sequence of the scheme, blocks transduction ; and (ii) the onset (i.e. transduction) time of agonist responses is shortened when the cells are pretreated with PMA or OAG, which act on protein kinase C and short-cut the rate-limiting sequence.

References

Babior B.M. (1978). Oxygen-dependent microbial killing by phagocytes. **New Engl. J. Med.** **298** : 659-668.

Baggiolini M. (1984). Phagocytes use oxygen to kill bacteria. **Experientia 40** : 906-909.

Baggiolini M. and Dewald B. (1984). Exocytosis by Neutrophils. **Contemp. Top. Immunobiol. 14** : 221-246.

Baggiolini M., Dewald B. and Thelen M. (1988). Effects of PAF on neutrophils and mononuclear phagocytes. **Biochem. Pharmacol.** (Karger Series) **22** (in press).

Berridge M.J. (1984). Inositol trisphosphate and diacylglycerol as second messengers. **Biochem. J. 220** : 345-360.

Cockcroft S. and Gomperts B.D. (1985). Role of guanine nucleotide binding protein in the activation of polyphosphoinositide phosphodiesterase. **Nature 314** : 534-536.

Dewald B. and Baggiolini M. (1985). Activation of NADPH oxidase in human neutrophils. Synergism between fMLP and the neutrophil products PAF and LTB4. **Biochem. Biophys. Res. Commun. 128** : 297-304.

Dewald B., Payne T.G. and Baggiolini M. (1984). Activation of NADPH oxidase of human neutrophils. Potentiation of chemotactic peptide by a diacylglycerol. **Biochem. Biophys. Res. Commun. 125** : 367-373.

Fernandez H.N., Henson P.M., Otani A. and Hugli T.E. (1978).
    Evaluation of C3a and C5a leukotaxis in vitro and simula-
    ted in vivo conditions. J. Immunol. 120 : 109-115.
Ford-Hutchinson A.W., Bray M.A., Doig M.V., Shipley M.E. and
    Smith M.J.H. (1980). Leukotriene B, a potent chemokinetic
    and aggregating substance released from polymorphonuclear
    leukocytes. Nature 286 : 264-265.
Grzeskowiak M., Della Bianca V., Cassatella M.A. and Rossi F.
    (1986). Complete dissociation between the activation of
    phosphoinositide turnover and of NADPH oxidase by formyl-
    methionyl-leucyl-phenylalanine in human neutrophils deple-
    ted of Ca^{2+} and primed by subthreshold doses of phorbol
    12, myristate 13, acetate. Biochem. Biophys. Res. Commun.
    135 : 785-794.
Hamill O.P., Marty A., Neher E., Sakman B. and Sigworth F.J.
    (1981). Improved patch-clamp techniques for high resolu-
    tion current recording from cells and cell-free membrane
    patches. Pfluegers Arch. Ges. Physiol. 39 : 85-100.
Horn W. and Karnovsky M.I. (1986). Features of the transloca-
    tion of protein kinase C in neutrophils stimulated with
    the chemotactic peptide f-Met-Leu-Phe. Biochem. Biophys.
    Res. Commun. 139 : 1169-1175.
Ingraham L.M., Coates T.D., Allen J.M., Higgins C.P., Baehner
    R.L. and Boxer L.A. (1982). Metabolic, membrane, and
    functional responses of human polymorphonuclear leukocytes
    to platelet-activating factor. Blood 59 : 1259-1266.
Kaibuchi K., Takai Y., Sawamura M., Hoshijiama M., Fujikura T.
    and Nishisuka Y. (1983). Synergistic functions of protein
    phosphorylation and calcium mobilization in platelet
    activation. J. Biol. Chem. 258 : 6701-6704.
Krause K.H. and Lew D.P. (1987). Subcellular distribution of
    Ca^{2+} pumping sites in human neutrophils. J. Clin. Invest.
    80 : 107-116.

Moslki T.F.P., Naccache P.H., Marsh M.L., Kermode J., Becker
   E.L. and Sha'afi R.I. (1984). Pertussis toxin inhibits the
   rise in the intracellular concentration of free calcium
   that is induced by chemotactic factors in rabbit neutrophils : Possible role of the "G proteins" in calcium
   mobilization. **Biochem. Biophys. Res. Commun.** **124** :
   644-650.

Ohta H., Okajima F. and Ui M. (1985). Inhibition by islet-
   activating protein of a chemotactic peptide-induced early
   breakdown of inositol phospholipids and Ca^{2+} mobilization
   in guinea pig neutrophils. **J. Biol. Chem.** **260** : 15771-
   15780.

Okajima F., Katada T. and Ui M. (1985). Coupling of guanine
   nucleatide regulatory protein to chemotactic peptide
   receptors in neutrophil membranes and its uncoupling by
   islet-activating protein, pertussis toxin. **J. Biol. Chem.**
   260 : 6761-6768.

Peveri P., Walz A., Dewald A. and Baggiolini M. (1988). A
   novel neutrophil-activating factor produced by human
   mononuclear phagocytes. **J. Exp. Med.** (in press).

Prentki M., Wollheim C.B. and Lew P.D. (1984). Ca^{2+} homeos-
   tasis in permeabilized neutrophils : characterization of
   Ca^{2+} sequestering pools and the action of inositol 1, 4,
   5-triphosphate. **J. Biol. Chem.** **259** : 13777-13782.

Repine J.E., White J.G., Clawson C.C. and Holmes B.M. (1974).
   The influence of phorbol myristate acetate on oxygen
   consumption by polymorphonuclear leukocytes. **J. Lab. Clin.**
   Med. **83** : 6761-6768.

Showell H.J., Freer R.J., Zigmond S.H., Schiffmann E.,
   Aswanikumar S., Corcoran B. and Becker E.L. (1976). The
   structure-activity relations of synthetic peptides as
   chemotactic factors and inducers of lysosomal enzyme
   secretion for neutrophils. **J. Exp. Med.** **143** : 1154-1169.

Sklar L.A., Hyslop P.A., Oades Z.G., Omann G.M., Jesaitis A.J,
    Painter R.G. and Cochrane C.G. (1985). Signal transduction
    and ligand-receptor dynamics in the human neutrophil. J.
    Biol. Chem. 260 : 11461-11467.
Smith C.D., Cox C.C. and Snydermann R. (1985). Receptor-cou-
    pled activation of phosphoinositide-specific phospholipase
    C by an N protein. **Science 232** : 97-100.
Spaet A., Bradfort P.G., Mc Kinney J.S., Rubin R.P. and Putney
    Jr. J.W. (1986). A saturable receptor for $32p$-inositol-1,
    4,5-trisphosphate in hepatocytes and neutrophils. **Nature**
    319 : 514-516.
Von Tscharner V., Deranleau D.A. and Baggiolini M. (1986).
    Calcium fluxes and calcium buffering in human neutrophils.
    J. Biol. Chem. 261 : 10163-10168.
Von Tscharner V., Prodhom B., Baggiolini M. and Reuter H.
    (1986). Ion channels in human neutrophils activated by a
    rise in free cytosolic calcium concentration. **Nature 324** :
    369-372.
Walz A., Peveri P., Aschauer H. and Baggiolini M. (1987).
    Purification and amino acid sequencing of NAF, a novel
    neutrophil-activating factor produced by monocytes.
    Biochem. Biophys. Res. Commun. 149 : 755-761.
Wolf M., LeVine III H., Stratford May Jr W., Cuatrecasas P.,
    Sahyoun N. (1985). A model for intracellular translocation
    of protein kinase C involving synergism between Ca^{2+} and
    phorbol esters. **Nature 317** : 546-549.
Wymann M.P., Von Tscharner V., Deranleau D.A. and Baggiolini
    M. (1987). The onset of the respiratory burst in human
    neutrophils. **J. Biol. Chem. 262** : 12048-12053.
Wymann M.P., Von Tscharner V., Deranleau D.A. and Baggiolini
    M. (1987). Chemiluminescence detection of H_2O_2 produced by
    human neutrophils during the respiratory burst. **Anal.**
    Biochem. 165 : 371-378.

Yoshimura T., Matsushima K., Tanaka S., Robinson E.A., Appella
    E., Oppenheim J.J. and Leonard E.J. (1987). Purification
    of a human monocyte-derived neutrophil chemotactic
    factor that has peptide sequence similarity to other host
    defense cytokines. **Proc. Natl. Acad. Sci. USA 84** : 9233-
    9237.

MONOCLONAL ANTIBODIES FOR IMMUNOANALYSIS

M. DELAAGE & A. MOREL

Monoclonal antibodies have brought about a revolution in immunoanalysis by giving access to the whole antibody repertory. So far the only accessible usable antibodies were those corresponding to dominant antigens or dominant epitopes. Unlike a popular idea the problem with polyclonal antibodies was not their heterogeneity, but rather the stereotyped character of the response to complex antigens which precluded the emergence of antibodies to minor antigens.

Talking about repertory one should carefully distinguish :

- The antibody repertory defined in a given species by the genomic recombinations giving rise to the variable regions and then the antibody binding site.

- The antigen repertory consisting of three dimensional arrangements of atoms to which may correspond one or several antibodies, at a given level of affinity. The antigen repertory is less precisely defined and considerably more outspread than the antibody repertory, and actually not evaluable, even restricted to protein antigens.

The antibody repertory looks almost illimited for our investigation means, so that some theoricians did not hesitate to confuse them in a poetic theory of mirrors. Here we have to consider antibodies as an open system of recognition. The art in immunoanalysis is to seek the proper antibody to recognize a given antigen, and to look for new antigens recognized by a given antibody.

The antibody binding site is shared by both variable regions of heavy and light chain of the antibody. However it represents only a small area (300 $\overset{\circ}{A}{}^2$) of the variable domains (cross section 3000 $\overset{\circ}{A}{}^2$). The reason lies in the rule of stability of proteins. The minimum of free energy which defines the more stable conformation is obtained via specific interactions of residues with each other, or with the solvent if they are hydrophilic.

Few residues can be taken apart for their contribution to minimize free energy. They are accessible for ligand and constitute the active site, at the extremity of β fingers in the tertiary structure. These privilegiate residues are necessarly few.

The size of the antibody binding site defines that of the epitope. It does not exceed four aminoacid residues if the antigen is a protein or a peptide, that is equivalent to the size of a nucleotide, steroid or most of drugs. This is the reason why antibodies are more and more used for pharmacokinetics.

Having these features in mind we can now get into the specificity of monoclonal antibodies and immunoanalytic reactions. The term of specificity has several meanings depending on the object :

. For the biochemist the specificity is established considering a family of molecules, the steroides found in mammals for example. Specificity is defined by subclasses of molecules recognized at a given affinity level (10^{-9} M for example).

. For the analyst the binding specificity of a given ligand is often appreciated in a complex medium and must take into account the concentrations or affinities of the different possible ligands. If Ci and Ki are the concentration and dissociation constants of the different ligands respectively,

the titration of the antibody is described by a partition
function :

$$Z = 1 + \Sigma Ci/Ki$$

and the specificity is represented by the ratio between the
concentration of "specific" complex and other complexes :

$$\frac{Ci/Ki}{Z - 1 - Ci/Ki}$$

Other definitions can be written for more complex
situations involving several antibodies, for example sandwich
reactions.

The antigen specificity has another distinct outlook
in immunoanalysis. Although quite different it is often
referred as antibody specificity. The problem lies on the
representativity of epitopes. If we consider a tetrapeptide,
the number of combination of 4 a.a. is 20^4 = 160.000. Then,
the probability of finding the same tetrapeptide in two
independent proteins in a given species is far from being
negligible. For example the pentapeptide TyrGlyGlyPheMet of
enkephalin is found in three different independent proteins
(Dayhoff et al., 1978). The frequency of such events is rather
high because of gene duplication. Fortunately the conformational effects reduce this probability.

From this side the situation is more favourable with
polyclonal antibodies when used in precipitation reaction. In
this case the coincidence of several epitopes is required, and
the probability of unexpected cross-reaction is much less
frequent. On the opposite for immunobloting reactions or
immunohistochemistry, the heterogeneity of the polyclonal
antibodies increases the probability of cross-reaction.

The representativity of antigens may also define specificity. A same protein can occur in different cell types. This is well known for many enzymes and cytosqueleton proteins which have an ubiquitous distribution or glycosylated epitopes like blood group antigens which are largely distributed among a variety of normal and tumoral cells.

Some proteins are restricted to cells having a same embryonic origin. The S-100 protein initially found in nervous tissue has been identified in exocrine sudoral cells and melanocytes, and more generally in tumors of neuronal origin (Kanitakis & Thivolet, 1987). Neurone specific enolase is also a very common antigen in neuronal cells and tumors (small cells lung cancer (Kanitakis & Thivolet, 1987)).

In some cases a protein may be associated with a family of cell types having similar function. For example villin (Robine et al., 1985), discovered in epithelial intestinal cells is characteristic of brush border epithelia and found in venal microtubule and in gall bladder.

The CD1 antigen initially found on thymocytes was then found on epidermic Langerans cell (Kanitakis & Thivolet, 1987).

The question of affinity needs some comments. As the antibody site is limited, so is limited the epitope, and limited the affinity in the range 10^{-9} - 10^{-12} M. In this respect monoclonal antibodies have not modified the situation. The affinity is a flexible parameter which can be modulated by the operator, like the "stringency" of hybridation.

The Kd value is the ratio of the rate constants kd/ka. When considering different systems within a large range of affinity ka varies in rather narrow limits whereas kd is

highly variable and reflects the whole range of affinity. ka
reflects the probability of an efficient encounter, kd
reflects the half-life of the complex :

$$kd = \frac{0.7}{t_{1/2}}$$

The affinity can be modulated by pH, temperature,
solvent and other components in large limits (at least 10^8)
and this is used in immunopurification systems. It can also be
modulated by chemical derivatization of the antigen as shown
with histamine. Histamine is such a small molecule that it
cannot fill properly an antibody binding site. The attempts to
raise antibodies recognizing directly histamine never gave
affinity better than 10^{-6} M. Since histamine has to be coupled
to a carrier to be made immunogenic we chose to also transform
histamine in the sample with the coupling agent. Thus modified
histamine resembles the immunogen and is more immunoreactive.
The link, succinylglycinamide was especially designed to
fullfill a set of technical and industrial requirements. The
rationale of the system is presented in Table I.

Table I : Rationale of histamine radioimmunoassay

Immunogen	: histamine-succinyl-glycyl-albumine
Tracer	: histamine-succinyl-glycyl-tyrosinamide (^{125}I)
Reagent	: NHS-succinyl-glycinamide
Acylated histamine	: histamine succinyl-glycinamide

The iodinated derivative was designed with the same
structure as immunogen, i.e. with the succinyl glycinamide
group on the side chain. A monoclonal antibody has been
selected using this iodinated derivative. Histamine is not
assayed as such but after derivatization with succinylglycinamide N-hydroxysuccinimide ester which brings about an affinity

enhancement of 5×10^5 (Table II). The ligand histamine-succi-
nyl glycinamide is ionisable and hydrophilic so that its
affinity for antibody depends strongly on temperature and pH.
Only the uncharged form of imidazole is properly recognized by
the monoclonal antibody. Histamine hemisuccinate (and other
derivatives bearing a carboxylic group) is very poorly
recognized, merely 5 times better than unmodified histamine.
The carboxylate is likely to form a salt bridge with the
cationic form of imidazole, thus stabilizing a non immunoreactive conformation of the hapten.

The specificity of the binding is excellent considering the possible cross-reactions. The relative affinities of
histidine (precursor of histamine) and 3 Me-histamine (the
first degradation product) are below $1/10^5$ and $1/10^4$ respectively.

Table II : Specificity of anti histamine antibodies

Histamine succinyl glycinamide	Kd = 1.1×10^{-10}	M
3 Me Histamine succinyl glycinamide	1.6×10^{-6}	M
Histidine succinyl glycinamide	2.8×10^{-5}	M
Histamine hemi succinate	1.1×10^{-5}	M
Histamine	5.5×10^{-5}	M

Accordingly histamine can be assayed directly, without extraction or concentration from plasma, or any biological
fluid or cell supernatant. One of the main application of the
histamine immunoassay is the histamine release by basophils
(figure 1), triggered by the cross-link of surface IgE by
allergen (Lichtenstein & Osler, 1964). A recent improvment of
this method (Weyer et al., 1986) consist in using washed
basophils from a healthy subject, to test the IgE of patients
with calibrated allergens.

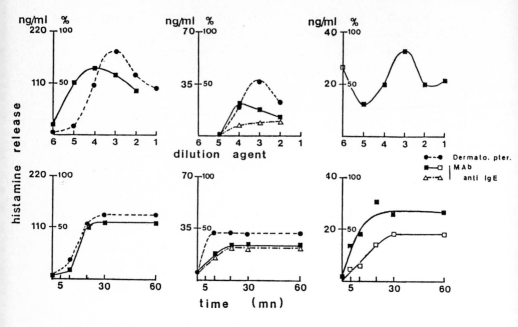

Figure 1 : Histamine release from whole blood

    a - Diluted whole blood from three different patients was incubated with various amounts of allergen or monoclonal anti IgE for 30 min at 37°C (upper part).

    b - The concentration which gave maximum histamine release was incubated for various times at 37°C using the same blood samples (lower part).

Monoclonal antibodies are powerfull tools for immunoanalysis. Their specificities are currently used for cell
typing or protein identification. Even for haptens they allow
the access to the very specific antibody which shows the
required affinity.

References

Dayhoff M.O., Hunt L.J., Barker W.C., Schwartz R.M., Orcutt
B.C. (1978). **Protein segment dictionary.**

Kanitakis J., Thivolet J. (1987). Les immunomarquages en
histologie cutanée. **Ann. Pathol. 7** : 79-97.

Lichtenstein L.M., Osler A.G. (1964). Studies on the mechanisms of hypersensitivity phenomena. **J. Exp. Med.** :
507-530.

Robine S., Huet C., Mole R., Sahuquillo-Merino C., Coudririer
E., Zweibaum A., Louvard D. (1985). Can villin be used to
identify malignant and undifferentiated normal digestive
epithelial cells ? **Proc. Natl. Acad. Sci. USA 82** :
8488-8492.

Weyer A., Ougen P., Dandeu J.P., Marchand F., David B. (1986).
Passive "in vitro" sensitization of human basophils with
dermatogoides farinae specific IgE. **Agents Actions 18** :
178-181.

PART TWO : EXPERIMENTAL SECTION

Experiment n° 1

TRANS-MEMBRANE SIGNALING VIA PROTEIN KINASE C
AND ITS INHIBITION BY STAUROSPORINE

C.W. MAHONEY, R. LUTHY and A. AZZI

I. INTRODUCTION AND AIMS

Protein kinase C (PKC) is thought to play an important role in the signal transduction across the cell membrane in different cell types (Nishizuka, 1984). Many different activators act **via** this pathway. PKC is activated by diacylglycerols that are produced by a phospholipase C as a consequence of the binding of various ligands to various receptors. The activated PKC then phosphorylates intracellular substrates which lead to the responses of the cell to the extracellular activator (Figure 1). On the other hand purified PKC binds phorbol esters, substances which have been known for many years as tumor promoters and non-physiological cell activators (Hecker and Schmidt, 1974). PKC is a Ca^{2+}-activated, phospholipid dependent enzyme and its activity is modulated by phorbol esters and diacylglycerols : they activate PKC by lowering the amount of Ca^{2+} and phosphatidylserine needed for maximal activity (Castagna et al., 1982). The binding of radioactive phorbol esters can be inhibited competitively by diacylglycerols (Sharkey et al. 1984, 1985), suggesting that diacylglycerol and phorbol esters have the same binding site.

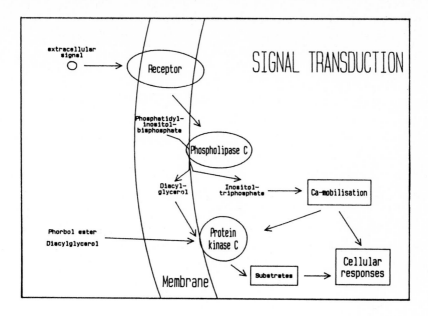

Figure 1 : Signal transduction

Recently Tamaoki et al. (1986) have reported that the antifungal agent staurosporine (Fig.2) is a potent inhibitor of purified PKC. Staurosporine, a metabolite secreted by **Streptomyces sp.** and first isolated by Omura et al. (1977), does not compete for any of the other ligands of PKC (i.e. Ca^{2+}, histone, DAG, ATP, PDB, and PS) and yet is the most potent inhibitor of PKC (IC_{50} = 2.7 nM) isolated to date (Tamaoki et al. 1986). This potent inhibitory action suggests the use of staurosporine in cell studies in order to dissociate PKC effects from other cell activation processes. Because staurosporine also inhibits purified cAMP dependent protein kinase (Tamaoki et al., 1986) careful controls are required in cell studies to rule out other interactions of the inhibitor with the **in vivo** system. In the **in vivo** experiments described here we assume that the primary site of action for staurosporine is on PKC. We demonstrate here the **in vitro**

inhibition of purified PKC, both Ca^{+2}- PS- activated and PDB-
PS- activated forms (Fig. 4 & 5), and the **in vivo** inhibition
of PDB- activated platelet aggregation (Fig. 6) by staurosporine.

* <u>Abbreviations</u> : PS, phosphatidylserine ; EGTA, ethylenegly-
colbis-(2-aminoethyl)-tetraacetic acid ; EDTA, ethylenediami-
notetraacetic acid ; ATP, adenosine-5'-triphosphate ; TCA,
trichloroacetic acid ; NaPPi, sodium pyrophosphate ; BSA,
bovine serum albumin ; PKC, protein kinase C ; DAG, diacylglycerol ; PDB, phorbol-12, 13-dibutyrate ; PEG, polyethylenegly-
col ; TLC, thin layer chromatography ; CBB-G, Coomassie
brilliant bule G ; DMSO, dimethylsulfoxide ; DMF, dimethyl-
formamide ; DW, doubly distilled water ; kD, kilodalton ; BME,
2-mercaptoethanol.

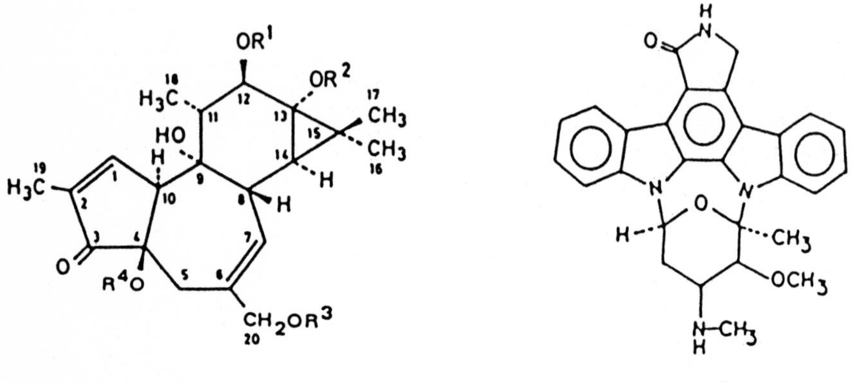

Figure 2 : Structures of phorbol 12, 13 dibutyrate (R_1 & R_2,
butyrate ; R_3 & R_4, H) (A) and staurosporine (B)

II. EQUIPMENT, CHEMICALS AND SOLUTIONS

A. Equipment

1. For purification of PKC from Bovine Brain

- Refrigerated preparative centrifuge
- Glass wool
- 100-200 ml teflon-glass homogenizer
- Dialysis tubing
- Phenyl Sepharose, 50 ml
- DEAE cellulose, 25 ml (Whatman DE 52)
- Columns, fraction collector, detector, gradient maker
- Waring blender

2. For phosphorylation assay

- Water bath at 30°C
- Vacuum manifold for 25 mm filter discs
- Millipore HA (0.45 μm) or Whatman GFB (1.0 μm) Filter discs
  (25 mm diameters)

3. For preparation of platelet rich plasma (PRP)

- Preparative centrifuge

4. For activation of platelets

- Thermostatted recording spectrophotometer
- Water bath at 37°C

B. Solutions

1. For purification of PKC from bovine brain

- Buffer H : 0.25 M sucrose, 20 mM Tris-HCl, 2 mM EDTA, 10 mM EGTA, 2 mM BME (pH 7.5)
- Buffer A : 20 mM Tris-HCl, 1 mM EDTA, 1 mM EGTA, 2 mM BME (pH 7.5)
- 200 mM $CaCl_2$
- 200 mM $MgCl_2$
- Buffer B : 20 mM Tris-HCl, 0.1 mM $CaCl_2$, 2 mM BME (pH 7.5)
- Buffer B with 1 M NaCl
- Buffer C : 20 mM Tris-HCl, 1 mM EGTA, 2 mM BME (pH 7.5)
- Buffer C with 0.4 M NaCl

2. For phosphorylation Assay

- 0.5 M Tris-HCl, pH 7.5 (30°C)
- 25 mM Mg acetate
- 0.25 mM $(\gamma\text{-}^{32}p)$ ATP (100,000 cpm/nmol, Amersham)
- 5 mg/ml histone (lysine rich fraction, Sigma Type III-S)
- 200 mM $CaCl_2$
- 200 mM $MgCl_2$
- 10 mg/ml phosphatidylserine (PS) in $CHCl_3$ (Lipid Products)
- 12 % trichloroacetic acid (TCA), 2 % sodium pyrophosphate (NaPPi) (4°C)
- 6 % trichloroacetic acid, 1 % sodium pyrophosphate (4°C)
- 10 mg/ml bovine serum albumin (BSA)
- PKC enzyme
- 0.5 mM ATP (a_{259} = 15,400 $M^{-1}cm^{-1}$)
- 20 mM Tris-HCl, pH 7.5 (30°C)
- 400 µM PDB in ethanol (Sigma)
- 0.2 mg/ml diolein in ethanol (optional) (Sigma)
- 0.67 µM staurosporine in 3.3 % DMF, 20 mM Tris-HCl (pH 7.5)

Final concentrations of assay mix (0.25 ml total volume) :

20 mM Tris-HCl, pH 7.5 (30°C)
5 mM Mg acetate
10 μM (γ-^{32}P) ATP (100,000 cpm/nmol)
0.2 mg/ml histone
0-500 μM CaCl$_2$
0-50 μg/ml PS
0-50 μM PDB
0-50 nM staurosporine
0-1.0 μg/ml diolein (optional)

3. For preparation of platelet rich plasma (PRP)

ACD :   Na citrate   75 mM
         Citric acid   38 mM
         Glucose 124 mM
Suspension buffer :
         HEPES   31 mM   pH 7.4
         Glucose   5 mM
         KCl   3 mM
         NaCl   118 mM

III. EXPERIMENTAL PROCEDURES

1. Purification of PKC from Bovine Brain (modified procedure of Walsh et al., 1984) (all steps at 4°C)

Homogenization

Fresh bovine brain (400 g) is homogenized in a Waring Blender (30 sec, twice) with 3 vols. of Buffer H. The homogenate is centrifuged (30 000 X g, 15 min) and the resulting supernatant is filtered through glass wool. The

filtrate is centrifuged (60 000 X g, 105 min.) and filtered
through glass wool and is then made 75 % (w/v) in ammonium
sulphate and stirred for 30 min. The precipitated protein is
then collected by centrifugation (30 000 X g, 45 min.). The
protein pellet is suspended in 100-200 ml buffer A and is
homogenized, and dialyzed extensively against buffer A.

Phenyl Sepharose chromatography

The dialyzed protein is clarified by centrifugation
(30 000 X g, 30 min) and is then made 2 mM in $CaCl_2$ and $MgCl_2$
and applied to a Phenyl Sepharose (1.5 x 30 cm) column
equilibrated in Buffer B. The column is washed with Buffer B,
followed by Buffer B, 1 M NaCl and Buffer C (3 bed volumes
each). PKC and < 5 % of the applied protein elute in the last
buffer. The PKC active fractions from the Buffer C eluate are
pooled and dialyzed against Buffer C.

DEAE cellulose chromatography

The dialyzed PKC active pool is applied to a DE 52
(1 X 30 cm) column equilibrated in Buffer C. The column is
washed with Buffer C (3 bed vols.) and then eluted with a
linear gradient of Buffer C to Buffer C with 0.4 M NaCl (200
ml each). The PKC activity elutes at approximately 0.1 M NaCl.
The PKC active fractions are pooled and dialyzed against
Buffer C and stored at 4°C. Typical specific activities of
50-200 nmol Pi/min mg protein are obtained.

2. Protein Assay (Bradford, 1976 ; Spector, 1978)

Concentrated dye solution :
Dissolve 200 mg Coomassie brilliant blue G (CBB- G) in
100 ml 95 % ethanol. While stirring add 200 ml conc. H_3PO_4 and
then 100 ml DW. Store at 4°C.

Working dye solution (good for 2 weeks) :

Dilute concentrated dye solution 1 : 4 with DW and refrigerate for 2 hrs. Filter the solution through Whatman 1 paper and store at 4°C.

Add the working dye solution to the protein sample, vortex, and read the absorbance at 595 nm within 2-15 min. Use the following volumes for the specified protein range needed :

Protein range	Sample volume	Dye volume
0.5-5 µg	0.1 ml	0.7 ml
1-10 µg	0.1 ml	1.0 ml
10-100 µg	0.1 ml	5.0 ml

Make a standard curve using a 1 mg/ml BSA solution $(a_{280} = 0.66 \ (ml/mg).cm^{-1})$.

3. Purity Analysis of Phosphatidylserine by TLC

Merck silica 60 or silica 60 F (5 x 20 cm) on glass TLC plates are used. Phosphatidylserine (1-100 µg loads) in $CHCl_3$ are spotted on the TLC plate 2.5 cm up from the bottom edge. The plate is air dried and then developed in $CHCl_3$/MeOH/ 28 % NH_3 (69 : 26 : 5, by vol.). The plate is air dried in a hood and the phospholipids are detected by placing the plate in a chamber with iodine crystals. Within several minutes yellow spots are detectable. PS and diolein have R_fs of 0.10 and 0.94 respectively under these conditions. The use of an alternative solvent system, benzene/diethyl ether/ethyl aceta- te/acetic acid (80 : 10 : 10 : 0.2), results in R_fs of 0 and 0.58 for PS and diolein respectively. High purity PS is obtained from Avanti Polar Lipids (Birmingham, AL, USA) or Lipid Products (South Nutfeld, Surrey, UK).

4. Staurosporine Isolation and Purity Analysis

Staurosporine is isolated from the culture supernatant of **Streptomyces actuosus** or **Streptomyces staurosporeus** (Morioka et al., 1985 ; Omura et al., 1977). Purity is ascertained by TLC analysis on silica gel (CHCl$_3$: MeOH, 10 : 1, R$_f$: 0.55 ; benzene : acetone, 1 : 2, R$_f$: 0.43 ; butanol : acetic acid : DW, 8 : 1 : 1 ; R$_f$: 0.24) using iodine vapors for detection. The molar absorptivity for staurosporine is 57 300 M^{-1} cm^{-1} at 292 nm.

5. Phosphorylation Assay
9.00 am ⟶ 12.00

The amount of PS needed for an experiment is dispensed into a glass tube and dried down with a stream of N$_2$. To this is added 50 µl 20 mM Tris-HCl, pH 7.5 per assay and the mixture is sonicated at high power under N$_2$ and on ice (30 sec, twice). The sonicated mixture is then added to the balance of the assay mix. A stock solution of (γ-^{32}P) ATP (0.25 mM ; 100,000 cpm/nmol) is made by combining 0.5 vol 0.5 mM ATP (cold) with 0.489 vol DW and 0.0114 vol 1 mCi/ml (γ-^{32}P) ATP (Amersham PB.132 in ethanol/DW 1 : 1) and can be stored frozen in small aliquots. Radioactive work must be done behind plexiglass shielding and with the use of gloves, lab coat, and safety glasses to protect against the hard radiation of ^{32}P.

To start the reaction 20 µl of PKC enzyme is added to 230 µl of assay mix and the tube is gently mixed and incubated at 30°C for 8 min. The reaction is stopped by the addition of 0.35 ml cold 12 % TCA, 2 % NaPPi and 100 µl 10 mg/ml BSA. After mixing the contents are transferred quantitatively to an HA or GFB filter on a vacuum manifold. The empty tube is washed with cold 6 % TCA, 1 % NaPPi and the contents transferred to the filter. The filter is then extensively

washed (4-5 portions, several ml each) with cold 6 % TCA, 1 %
NaPPi and the filter is transferred to a minivial to which
5 ml of scintillation cocktail is added. After wiping the
vials with a wet Kimwipe (eliminates potential electrostatic
artifactual counts), the minivials are counted in a liquid
scintillation counter.

　　　Phosphorylation specific activities can be determined
by performing the above assay on a time course basis (0-8 min)
and by measuring the protein content of the PKC source.

　　　To ascertain the presence of Ca^{2+} - and phospholipid-
dependent protein kinase (versus other protein kinases) the
assay should be done in the presence of 0.5 mM $CaCl_2$ alone,
0.04 mg/ml PS alone, 0.5 mM $CaCl_2$ and 0.04 mg/ml PS together,
and in the absence of both $CaCl_2$ and PS. Maximal activity for
PKC should be obtained only in the presence of both Ca^{2+} and
PS (Figure 3).

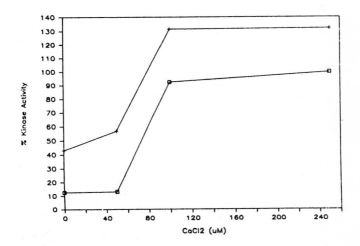

Figure 3 : Ca^{2+} dependence of PDB- and control-activation
(absence of PDB) of partially purified PKC. PDB (50 nm, +-+),
control (absence of PDB, open squares). All assays were
performed in the presence of 10 µg/ml PS

PKC binds and is stimulated by diacylglycerols (DAG) and phorbol diesters in addition to PS. Diacylglycerols and phorbol diesters lower the free Ca^{2+} concentration necessary to get maximal stimulation and elevate the kinase activity above the Ca+- PS- maximally ·activated levels (Kikkawa et al., 1983 ; Mori et al., 1982). Ca^{2+}- PS- and PDB- PS- Ca^{2+}- activation of the kinase activity may be plotted versus the $CaCl_2$ concentration (Fig. 3). These activation curves may alternatively be normalized to each curves maximum activity to calculate the $CaCl_2$ $_{50}$ (the concentration of $CaCl_2$ required for 0.5 maximal activity). Inhibition of kinase activity by staurosporine data may be plotted as a function of staurosporine or $CaCl_2$ concentration (Figs. 4 & 5) and the IC_{50} values (concentration of staurosporine that gives 50 % inhibition of kinase activity) may be calculated.

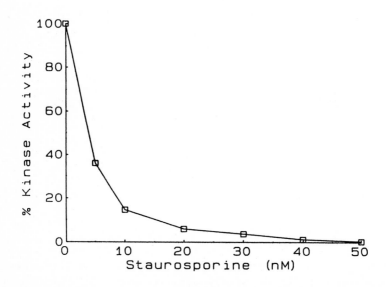

Figure 4 : Concentration dependence of the inhibition of partially purified PKC by staurosporine. All assays were carried out in the presence of 500 μM $CaCl_2$ and 40 μg/ml PS

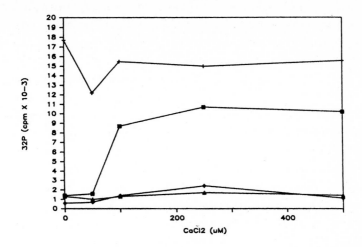

Figure 5 : The lack of deinhibition by PDB (50 μM) of staurosporine (50 nM) inhibited partially purified PKC. All assays are in the presence of 10 μg/ml PS. Control (absence of PDB, squares) ; 50 μM PDB activated (+-+) ; 50 nM staurosporine inhibited control (absence of PDB, diamonds) ; 50 nM staurosporine inhibited, 50 μM PDB activated (triangles)

6. Preparation of platelet rich plasma (PRP)

2.00 pm ⟶ 4.00 pm

Since blood platelets can be activated by exposure to glass, they should be centrifuged and stored in plastic tubes. The starting material can either be anticoagulated blood or a buffy coat. ACD is used most often as an anticoagulant at a ratio of blood to AC of 5 : 1.

Principle

Platelets activated with PDB form aggregates. The formation of these aggregates can be followed by the decrease of the absortion of light because the initial suspension of platelets scatters more light than the aggregates. The rate of aggregation is obtained from the maximal slope of the time-resolved aggregation curves. Since phorbol dibutyrate specifically and directly activates protein kinase C an inhibitor of this kinase should also inhibit the activation of platelets by PDB.

Procedure

         Platelets (2 ml) preincubated at 37°C are put in a plastic cuvette and placed in the holder of the spectrophoto- meter. The suspension is stirred and the recording of the absorbance at 480 nm is started. The staurosporine is added (concentrations from 0 to 100 nM) and the sample is incubated for 2 min. Then 20 µl of the CaCl$_2$ stock solution are added (calcium is necessary for aggregation). After one minute the aggregation is started by the addition of 5 µl PDB solution (Fig. 6). The changes is transmission are recorded and the maximal slope is determined and plotted vs. the concentration of staurosporine.

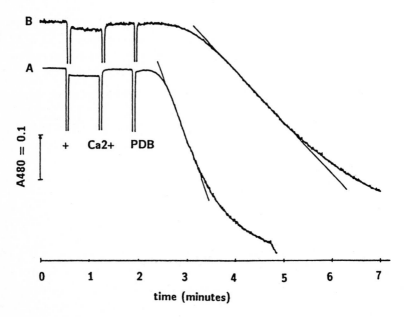

Figure 6 : Time-courses of platelet aggregation stimulated with 1 µM PDB. +, addition of solvent for control (A) or of 20 nM staurosporine (B)

IV. REFERENCES

Bradford M. (1976). A rapid and sensitive method for the quantitation of microgram quantities of protein utilizing the principle of protein-dye binding. **Anal. Biochem. 72** : 248-255.

Castagna M., Takai Y., Kaibuchi K., Sano K., Kikkawa U., Nishizuka Y. (1982). Activation of protein kinase C by phorbol esters. **J. Biol. Chem. 257** : 7847-7851.

Hecker E., Schmidt R. (1979). Phorbol esters - the irritants and cocarcinogens of **Croton Tiglium** L. **Fortscher. Chem. Org. Naturstoff. 31** : 377-467.

Kikkawa U., Minakuchi R., Takai Y., Nishizuka Y. (1983). Calcium-activated, phospholipid-dependent protein kinase (protein kinase C) from rat brain. **Methods Enzymol. 99** : 288-298.

Morioka H., Ishihara M., Shibai H., Suzuki T. (1985). Staurosporine induced differentiation in a human neuroblastoma cell line, NB-1. **Agric. Biol. Chem. 49** : 1959-1963.

Nishizuka Y. (1984). The role of protein kinase C in cell surface signal transduction and tumor promotion. **Nature 308** : 693-698.

Omura S., Iwai Y., Hirano A., Nakagawa A., Awaya A., Tsuchiya H., Takahashi Y., Masuma R. (1977). A new alkaloid AM-2282 of **Streptomyces** origin taxonomy, fermentation, isolation and preliminary characterization. **J. Antibiotics 30** : 275-282.

Sharkey N.A., Blumberg P.M. (1985). Kinetic evidence that 1,2-diolein inhibits phorbol ester binding to protein kinase C via a competitive mechanism. **Biochem. Biophys. Res. Comm. 133** : 1051-1056.

Sharkey N.A., Leach K.L., Blumberg P.M. (1984). Competitive inhibition by diacylglycerol of specific phorbol ester binding. **Proc. Natl. Acad. Sci. 81** : 607-610.

Spector T. (1978). Refinement of the coomassie blue method of
    protein quantitation. **Anal. Biochem. 86** : 142-146.
Tamaoki T., Nomoto H., Takahashi I., Kato Y., Morimoto M.,
    Tomita F. (1986). Staurosporine, a potent inhibitor of
    phospholipid/Ca^{++}-dependent protein kinase. **Biochem.
    Biophys. Res. Commun. 135** : 397-402.
Walsh M.P., Valentine K.A., Ngai P.K., Carruthers C.A.,
    Hollenberg M.D. (1984). Ca^{2+}-dependent hydrophobic-
    interaction chromatography. **Biochem. J. 224** : 117-127.

Experiment n° 2

ENDOCYTOSIS ANALYSIS BY FLOW CYTOFLUOROMETRY : STUDY OF THE
ENDOCYTOSIS OF FLUORESCEINYLATED NEOGLYCOPROTEINS VIA
MEMBRANE LECTINS OF TUMOR CELLS

P. MIDOUX, A.C. ROCHE & M. MONSIGNY

I. INTRODUCTION AND AIMS

Membrane lectins bind specifically glycoproteins and
neoglycoproteins bearing appropriate mono or oligosaccharides.
These endogenous lectins have been characterized at the cell
surface of both normal and tumor cells (Monsigny et al.,
1983). Carbohydrate specific uptake of glycoproteins and
neoglycoproteins occur in normal cells such as hepatocytes,
Kupffer cells, macrophages, endothelial cells, fibroblasts and
several tumor cells (Monsigny et al., 1983 ; Ashwell &
Harford, 1982). Recently, neoglycoproteins have been shown to
be suitable to target cytotoxic drugs (Monsigny et al., 1983,
1984 a), gelonine (Roche et al., 1983), antiviral drugs (Fiume
et al., 1982) and immunomodulators (Monsigny et al. 1984 b).

In the cases of drug targetting it is important to
determine the endocytosis and the extent of the degradation of
the carriers inside the cells.

During receptor-mediated endocytosis, ligands are recognized by specific receptors on the surface of the cells, brought into the cells via endocytic vesicles derived from the plasma membrane. Although the exact pathway followed by the ligand is not known, the key step is a rapid acidification of endocytotic vesicles allowing the dissociation of ligands from their receptors. The fate of internalized material markedly differs from one to another ligand, but when internalized ligand-receptor complexes are usually dissociated, and ligands may then be delivered to lysosomes where proteolysis occurs (for reviews see Steinman et al., 1983 and Wileman et al., 1985).

Recently we described rapid flow cytofluorometric methods allowing the quantitative determination of the binding, the uptake and the degradation of an endocytosed fluoresceinylated protein ligand (Midoux et al., 1986, 1987). This method is based on the fluorescence properties of fluorescein-labeled ligands and on some properties of cell compartments : i.e., the lumen of endosomes and lysosomes is acidic ; the fluorescence of fluorescein bound to a protein is lower than that of free fluorescein ; the fluorescence of the ligand depends upon the number of fluorophore molecules bound to a protein (fig. 1) and upon its pH environment.

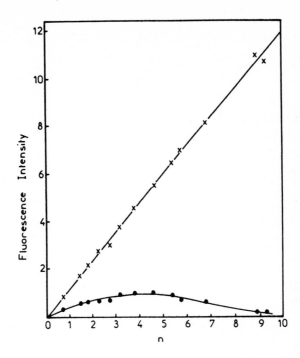

Figure 1 : Fluorescence properties of fluorescein bound to neoglycoproteins.

(● - ●) fluorescence intensity of 1 μg fluorescein-labeled neoglycoprotein in 1 ml PBS. (X-X) fluorescence intensity of 1 μg fluorescein-labeled neoglycoprotein in 1 ml PBS upon proteolytic digestion by pronase. n was average number of fluorescein residues bound to a neoglycoprotein molecule. In this study fluorecein-labeled neoglycoproteins contained a similar number of sugar residues per molecule (23 $\pm$ 6). The fluorescence intensities (excitation wavelength : 495 $\pm$ 7 nm; emission wavelength : 520 $\pm$ 7 nm) were expressed relatively to the fluorescence intensities of 2.55 x 10^{-7} M solution of quinine sulfate in 0.1 N H_2SO_4 (excitation wavelength : 345 $\pm$ 7 nm ; emission wavelength : 460 $\pm$ 7 nm). Reproduced with permission from Midoux et al., 1987.

II. EQUIPMENT, CHEMICALS AND SOLUTIONS

A. Equipment

1. Access to

- CO_2 incubator 37°C (5 % CO_2, 95 % air)
- Refrigerate centrifuge
- Ice

2. On the bench

- Becton Dickinson FACS Analyser
- Spectrofluorimeter
- Vortex
- Rubber policeman
- Plastic tubes
- Sterile Petri dishes
- Sterile culture plates (24 wells)
- Sterile centrifuge tubes (25 ml)

B. Chemicals

- Polystyrene sulfonate beads (Dionex DC-4A beads 10 μm $\pm$ 2 in diameter) (Dionex, Sunnyvale, CA, USA) containing given amounts of 1-(fluoresceinyl/thioureido)-4,8-diazaeicosane prepared according to Monsigny et al., 1984 a.
- Fluoresceinylthiocarbamyl neoglycoproteins F-BSA and α-Glc-F-BSA prepared according to Monsigny et al., 1984 a.
- Leupeptin, a bacterial tripeptide (N-acetyl-L-leucyl-L-leucyl-L-arginal) acting as a proteinase inhibitor from Sigma (St Louis, USA).
- Monensin, obtained from Calbiochem (La Jolla, USA) and a stock ethanolic solution (25 mM ; 17,5 mg/ml) prepared just before used.
- Bovine serum albumin (fraction V, Sigma).

- Fetal bovine serum (FCS) (Gibco, Renfrewshire, U.K.).
- Pronase : a PBS solution (1 mg/ml) of non specific proteases isolated from **Steptomyces griseus**, Pronase B grade 45000 PUK/g (Calbiochem, La Jolla, USA).

C. Solutions

Complete PBS containing 1 % BSA

NaCl	8 g/l
KCl	0,2 g/l
Na_2HPO_4, 12 H_2O	2,89 g/l
KH_2PO_4	0,2 g/l
$MgCl_2$, 6 H_2O	0,1 g/l
$CaCl_2$	0,1 g/l
BSA fraction V	10 g/l

Complete medium : Minimal essential medium (MEM) with Hank's salts containing 20 % heat inactivated FCS and 0.2 % glutamine.

Sheath fluid

NaCl	7,83 g/l
Na_2 EDTA	0,36 g/l
KCl	0,28 g/l
$KH_2 PO_4$	0,26 g/l
$Na_2 HPO_4$	2,35 g/l
NaF	0,64 g/l
2-phenoxyethanol	0,2 % (v/v)

Solution of quinine sulfate : 2.55 x 10^{-7} M in 0,1 N H_2SO_4

III. EXPERIMENTAL PROCEDURES

The FACS analyzer (flow cytofluorometer) is used to study the endocytosis of fluoresceinylated α-glucosylated bovine serum albumin which binds a sugar-binding receptor (membrane lectin) present at the surface of Lewis lung carcinoma (3LL) cells (Roche et al., 1983).

1. Studies of the binding and the uptake of α-Glc-FBSA by 3LL cells
9.00 am ⟶ 12.00 am

3LL cells (2 x 10^4 cells/ml) were grown in tissue culture plates (16 mm diameter) in minimal essential medium (MEM) with Hank's salts containing 20 % heat inactivated fetal calf serum (FCS) and 0.2 % glutamine (complete medium). The cells were maintained at 37°C in a humidified atmosphere of 5 % CO_2-95 % air. Non-adherent cells from 2 days cultured cells were removed and the adherent cells (1-2 x 10^5 cells) were incubated at either 4°C or 37°C for 2 hours in complete PBS containing fluorescein-labeled neoglycoproteins (α-Glc-F-BSA and FBSA). Supernatants were removed and then, the cells were harvested with a rubber policeman, washed twice in cold PBS plus 0.5 % BSA to remove unbound ligand and resuspended in cold sheath fluid. The cells fluorescence intensities were analysed at 4°C using a FACS analyzer (Becton Dickinson, Sunnyvale, CA, USA) before and after a post treatment with monensin (30 minutes at 4°C in the presence of 50 μM monensin). The size and the fluorescence intensity of each cell were simultaneously recorded at a rate of 300 cells/sec (excitation wavelength : 485 $\pm$ 10 nm ; emission wavelength : 530 $\pm$ 15 nm). The standardization of the flow cytofluorometer was achieved by using polystyrene sulfonate beads, (Dionex DC-4A beads ; 10 μm $\pm$ 2 in diameter) (Dionex, Sunnyvale, USA) containing various amounts of 1-(fluoresceinylthioureido)-4,8-dia-za-eicosane (Monsigny et al. 1984 a).

Determination of the amount of cell-associated fluoresceinylated neoglycoproteins

The mean mount (Q) of fluoresceinylated neoglycoprotein associated to a single labeled cell was determined by using the following equation (I) :

$$(I) \qquad Q = \frac{F}{F(N)} \cdot \frac{d}{b}$$

expressed as pg of fluorescein-labeled neoglycoprotein per cell, where F was the cell fluorescence intensity of one labeled cell incubated at 4°C minus the cell autofluorescence and measured with a cytofluorometer, F(N) was the fluorescence intensity of 1 µg fluorescein-labeled neoglycoprotein in 1 ml PBS measured with a spectrofluorometer, b was the bead fluorescence intensity of one labeled bead minus the bead autofluorescence measured with a cytofluorometer, d was the fluorescence intensity of 10^6 fluorescein-labeled beads in 1 ml PBS measured with a spectrofluorimeter. The fluorescence intensities F(N) and d were expressed relatively to the fluorescence intensity of a solution of quinine sulfate $(2.55 \times 10^{-7}$ M in 0.1 N H_2SO_4 ; excitation wavelength : 345 ± 7 nm ; emission wavelength : 460 ± 7 nm).

2. Studies of the degradation of endocytosed material
9.00 am. ⟶ 5.30 pm

2.1. By using a lysosomal, hydrolase inhibitor (leupeptin).

3LL cells were incubated at 4°C for 90 minutes in the presence of 100 µg/ml α-Glc-FBSA containing 7 fluorescein residues per protein molecule and then the cells were washed. The cells were warmed up to 37°C in the absence of α-Glc-FBSA and either in the presence of in the absence of 200 µM

leupeptin. The time course of the cell associated fluorescence
was monitored by flow cytofluorometry before and after a
treatment at 4°C with monensin.

2.2. By using two α-Glc-BSA containing different
number of fluorescein residue per protein molecule.

3LL cells were incubated at 37°C for 1, 2, 4 and 6
hours, in the presence of α-Glc BSA substituted either with 7
fluorescein residues (F_7-α-Glc-BSA) or with 3.5 fluorescein
residues ($F_{3.5}$-α-Glc-BSA). The cells were then analysed by
flow cytofluorometry before and after a treatment at 4°C with
monensin.

Determination of the amount of degraded and undegraded
fluoresceinylated neoglycoprotein associated to one cell

The cell fluorescence intensity F(M) (fluorescence
intensity of cells upon monensin treatment minus the cell
autofluorescence) of a single cell incubated at 37°C in the
presence of fluorescein-labeled neoglycoprotein postreated
with monensin and measured by cytofluorometry treatment is
expressed as the sum of the fluorescence intensities of
undegraded and degraded fluorescein-labeled neoglycoprotein
according to (II).

$$(\text{II} \qquad F(M) = [Q(N).F(N) + Q(NP).F(NP) \ ] \ . \ \frac{b}{d}$$

where F(N) and F(NP) were the fluorescence intensities of
1 µg/ml undegraded and degraded fluorescein-labeled neoglycoprotein measured with a spectrofluorimeter, respectively. Q(N)
and Q(NP) were the mean amounts of undegraded and degraded
fluorescein-labeled neoglycoprotein expressed in pg per cell,
respectively ; b and d were as defined for the equation (I).

The mean amount Q(NP) of fluorescein-labeled neoglycoprotein which was not degraded in the cell could be determined by using two neoglycoproteins containing different number of fluorescein residues per protein molecule and indexed 1 and 2, respectively. Q(NP) and Q(N) were determined by resolving the following linear equation system (III) :

$$F(M)_1 = [Q(N).F(N)_1 + Q(NP).F(NP)_1] \cdot \frac{b}{d}$$

(III)

$$F(M)_2 = [Q(N).F(N)_2 + Q(NP).F(NP)_2] \cdot \frac{b}{d}$$

IV. COMMENTS

The fluorescence properties of the fluorescein residues bound to a protein are used to analyse by flow cytofluorometry the neoglycoproteins endocytosis mediated by membrane lectins of Lewis lung carcinoma cells (3LL cells). The binding of fluorescein-labeled α-glucosylated serum albumin to 3LL cells at 4°C can easily be determined by flow cytofluorometry because under these conditions the environmental pH is neutral and the neoglycoprotein is not degraded. When the cells are incubated at 37°C in the presence of a fluorescein-labeled neoglycoprotein, the fluorescence intensity of a cell is low because of the low pH of endosomes and lysosomes but is increased upon a post-incubation at 4°C in the presence of monensin, a proton/sodium ionophore.

The quantum yield of fluorescein bound to a protein is dependent upon the number of fluorophore molecules bound to a protein molecule and upon the pH of the environmental medium. The mean fluorescence intensity of a fluorescein molecule bound to a protein decreases when the number of fluorescein

residues per protein molecule increases ; however, after
proteolytic digestion the mean fluorescence intensity of a
fluorescein molecule is constant and equal to that of free
fluorescein.

The extent of the proteolytic digestion of an endocytosed neoglycoprotein can be assessed by comparing, upon a
monensin post-incubation at 4°C, the high cell-associated
fluorescence of cells incubated in the absence of leupeptin
(an inhibitor of lysosomal proteases) and the relatively low
fluorescence intensity of cells incubated in the presence of
leupeptin.

The extent of the degradation of the endocytosed
neoglycoprotein can be determined with a flow cytofluorometer
by using two neoglycoproteins containing either a low or a
high number of fluorescein residues per neoglycoprotein
molecule. The method takes into account the extent of
fluorescence quenching of fluorescein bound to a protein with
regards to the number of fluorescein molecules per protein
molecule.

V. REFERENCES

Ashwell G. and Harford J. (1982). Carbohydrate specific
    receptors of the liver. **Ann. Rev. Biochem. 51** : 531-554.
Fiume L., Busi C. and Mattioli A. (1982). Lactosaminated human
    serum albumin as hepatotropic drug carrier. Rate of uptake
    by mouse liver. **FEBS Letters, 146** : 42-46.
Midoux P., Roche A.C. and Monsigny M. (1986). Estimation of
    the degradation of endocytosed material by flow cytofluoro-
    metry using two neoglycoproteins containing different num-
    bers of fluorescein molecules. **Biol. Cell , 58** : 221-226.

Midoux P., Roche A.C. and Monsigny M. (1987). Quantitation of the binding, uptake and degradation of fluoresceinylated neoglycoproteins by flow cytometry. **Cytometry 8** : 327-334.

Monsigny M., Kiéda C. and Roche A.C. (1983). Membrane glycoproteins/glycolipids and membrane lectins as recognition signals in normal and malignant cells. **Biol. Cell**, 47 : 95-110.

Monsigny M., Roche A.C. and Midoux P. (1984 a). Uptake of neoglycoproteins via membrane lectin(s) of L1210 cells evidenced by quantitative flow cytofluorometry and drug targeting. **Biol. Cell**, 51 : 187-196.

Monsigny M., Roche A.C. and Bailly P. (1984 b). Tumoricidal activation of murine alveolar macrophages by muramyldipeptide substituted mannosylated serum albumin. **Biochem. Biophys. Res. Commun.**, 121 : 579-584.

Roche A.C., Barzilay M., Midoux P., Junqua S., Sharon N. and Monsigny M. (1983). Sugar specific endocytosis of glycoproteins by Lewis lung carcinoma cells. **J. Cell. Biochem.**, 22 : 131-140.

Steinman R.M., Mellman I.S., Muller W.A. and Cohn Z.A. (1983). Endocytosis and recycling of plasma membrane. **J. Cell. Biol.**, 96 : 1-27.

Wileman T., Harding C. and Stahl P. (1985). Receptor-mediated endocytosis. **Biochem. J.**, 232 : 1-14.

Experiment n° 3

NEUTROPHILS. RESPIRATORY BURST AND EXOCYTOSIS

B. DEWALD, J. DOUSSIERE, F. MOREL, M. BAGGIOLINI, P.V. VIGNAIS

I. INTRODUCTION AND AIMS

Respiratory burst

Upon stimulation with chemotactic agonists, phorbol esters or phagocytosable particles, the neutrophils suddenly increase their oxygen consumption. This phenomenon is called the respiratory burst. Depending on the conditions, oxygen consumption can rise as much as 50-fold over that of the unstimulated cells. The respiratory burst is not inhibited by cyanide, showing that it is not related to mitochondrial respiration. It is due to the activation of the NADPH-oxidase (which is inactive in resting cells), catalyzing the reduction of molecular oxygen to superoxide at the expense of NADPH

$$2\ O_2 + NADPH \longrightarrow 2\ O_2^{\overline{\cdot}} + NADP^+ + H^+$$

Superoxide is rapidly converted to hydrogen peroxide by dismutation

$$2\ O_2^{\overline{\cdot}} + 2\ H^+ \longrightarrow H_2O_2 + O_2$$

$NADP^+$ is reconverted to NADPH via the hexose monophosphate shunt.

The respiratory burst can thus be measured as the cyanide-insensitive consumption of oxygen and the formation of superoxide or hydrogen peroxide. A more indirect measurement is that based on the activity of the hexose monophosphate shunt.

Superoxide, hydrogen peroxide and products derived therefrom are essential in the killing of many microorganisms by the neutrophils and the mononuclear phagocytes.

Exocytosis

Exocytosis is the active release of preformed material, present in cytoplasmic storage organelles, into the extracellular space. It is a selective process which depends on the fusion of the membrane of the storage organelles with the plasma membrane, and which occurs without loss of cytosolic proteins. The release of storage macromolecules is of major importance for neutrophil function. Lytic enzymes, proteinases in particular, are required for intracellular digestion in phagosomes and, presumably, for diapedesis and chemotaxis. The same enzymes are involved in tissue damage at sites of inflammation. The assessment of exocytosis and of compounds or conditions that influence this process is therefore of interest to many investigators.

The method for the measurement of the respiratory burst and exocytosis described here apply to human neutrophils. Most of them, however, may be used with minor adjustments for neutrophils of other species and other granulocytes.

II. EQUIPMENT, CHEMICALS AND SOLUTIONS

A. Equipment

1. For preparation of neutrophils from buffy coats

- Refrigerated centrifuge

2. For respiratory burst

2.1. Oxygen uptake

- Oxygraph with Clark electrode
- Incubation chamber and circulating water bath
- Strip chart recorder

Before use the apparatus is calibrated to room air and 0 percent oxygen. The room air calibration is performed with PBS equilibrated at 37°C. The anaerobic calibration is then obtained by adding 1 to 2 mg of sodium dithionite.

2.2. Superoxide production

- Double-beam spectrophotometer
- Circulating water bath
- Recorder

2.3. Hydrogen peroxide production

- Spectrofluorimeter with a thermostatted cuvette holder
- Recorder
- Circulating water bath
- Vortex mixer

3. For exocytosis

3.1. Exocytosis experiment

- Shaking thermostatted water bath
- Refrigerated centrifuge

3.2. Assay of markers

3.2.1. Vitamin B $_{12}$ - binding protein (specific granules)
- Gamma Counter
- Centrifuge

3.2.2. β-Glucuronidase (azurophil granules)
- Spectrofluorimeter
- Shaking thermostatted water bath

3.2.3. Lactate dehydrogenase (cytosol)
- Spectrophotometer with recorder

B. Chemical and Solutions

1. <u>For preparation of neutrophils from buffy coats</u>

- Phosphate-buffered saline, 137 mM NaCl, 2.7 mM KCl, 8.1 mM Na_2HPO_4 and 1.5 mM KH_2PO_4, pH 7.4 (PBS)
- PBS containing 10 U/ml heparin
- Ficoll-Hypaque (e.g. Lympho-paque from Nyegaard, Oslo, Norway or Ficoll-Paque from Pharmacia, Uppsala, Sweden)
- Sodium citrate 3.2 %
- NaCl 154 mM containing 0.05 mM $CaCl_2$ (saline/low Ca^{2+})
- NH_4Cl 155 mM containing 10 mM EDTA, adjusted to pH 7.4

2. <u>For respiratory burst</u>

- PBS
- PBS containing 0.9 mM $CaCl_2$ and 0.49 mM $MgCl_2$ (PBS/Ca^{2+}, Mg^{2+})
- Glucose 1 M in PBS
- Potassium cyanide (KCN) 0.3 M in PBS

- Sodium azide (NaN$_3$) 0.3 M in PBS
- Cytochrome c 0.02 M in PBS
- Superoxide dismutase (SOD) 10 mg/ml in H$_2$O
- Phorbol 12-myristate 13-acetate (PMA) in 1 mg/ml DMSO
- N-formyl-methionyl-leucyl-phenylalanine (fMLP) 10 mM in DMSO
        Shortly before use a 1:100 dilution to 0.1 mM is
prepared in PBS
- Sodium dithionite (Na$_2$S$_2$O$_4$) cryst.

2.2. Superoxide production

- PBS
- Glucose 1 M in PBS
- PMA 1 mg/ml DMSO
- Cytochrome c (type VI from horse heart) 10 mM in H$_2$O
- Sodium deoxycholate 15 % in H$_2$O
- NADPH 50 mM in H$_2$O
- NADH 50 mM in H$_2$O
- SOD 10 mg/ml H$_2$O
- Catalase 4000 U/ml H$_2$O

2.3. Hydrogen peroxide production

- PBS/Ca^{2+}, Mg^{2+}
- Glucose 1 M in PBS
- p-Hydroxyphenylacetate (PHPA) 10 mg/ml H$_2$O, adjusted to pH
  7.4 with NaOH
- Horseradish peroxidase (POD) 8 mg/ml H$_2$O
- PMA 1 mg/ml in DMSO
- fMLP 10 mM in DMSO ; a 1:500 dilution in PBS is prepared
  before use
- SOD 10 mg/ml in PBS
- Catalase 4000 U/ml in H$_2$O
- Perhydrol (30 % H$_2$O$_2$). Shortly before use a 1:10000 dilution
in H$_2$O is prepared to obtain an approx. 1 mM solution. The

exact concentration is determined spectrophotometrically ($E_{240\ nm} = 40\ M^{-1}\ cm^{-1}$).

3. For exocytosis

3.1. Exocytosis experiment
- PBS/Ca^{2+}, Mg^{2+}
- fMLP 10 mM in DMSO
- Triton X 100 10 % in H_2O
- PMA 1 mg/ml in DMSO
- Ionophore A 23187 5 mM in DMSO
- Cytochalasin B 5 mg/ml DMSO

Shortly before use appropriate dilutions of the stock solutions in DMSO are prepared with PBS.

3.2. Assay of markers

3.2.1. Vitamin B $_{12}$-binding protein (specific granules)

- Potassium phosphate buffer 0.1 M, pH 7.5
- Cyano [^{57}Co] cobalamin 180-300 µCi/µg, 0.05 µg/ml

Prepare a 1:500 dilution in 0.1 M potassium phosphate buffer, pH 7.5 (0.05 ng/0.5 ml) shortly before use.

3.2.2. β-Glucuronidase (azurophil granules)

- Sodium acetate buffer 0.1 M, pH 4.0
- 4-methylumbelliferyl-beta-D-glucuronide 10 mM in 0.1 M sodium acetate buffer, pH 4.0
- 4-methylumbelliferone 5 mM in ethanol

A 1:100 dilution in buffer is prepared before use.

Stop solution : A buffer containing 0.05 M glycine and 5 mM EDTA, adjusted to pH 10.4 with 1 M NaOH.

3.2.3. Lactate dehydrogenase (cytosol)

- Sodium phosphate buffer 0.1 M, pH 7.5
- Sodium pyruvate 4 mM in H_2O
- NADH 0.8 mM in H_2O
- Triton X-100 0.1 % in H_2O

III. EXPERIMENTAL PROCEDURES

A. Preparation of neutrophils from buffy coats
9.00 am ⟶ 11.00 am

Experimental procedure :

Buffy coats are usually obtained from blood banks or transfusion centers. They are prepared by centrifuging anti-coagulated blood (up to 500 ml) in glass bottles or plastic bags at 2600 g for 10 min at 20° with the brake off. The plasma supernatant is aspirated or pressed out of the bag and the buffy coat, i.e., the upper cell layer (50-60 ml) is collected. It contains the white cells and platelets, many erythrocytes, and some residual plasma. The buffy coat is centrifuged at 800 g for 5 min at 20°. A platelet-containing supernatant, which is discarded, and a sediment are obtained. 8 ml of the top layer of this sediment are collected with a Pasteur pipette and diluted to 30 ml with PBS containing 10 U/ml of heparin. Two 15-ml portions are transferred into 50-ml centrifuge tubes on top of a 10-ml layer of Lympho-Paque to which 0.58 ml of 3.2 % sodium citrate had been added. The tubes are centrifuged at 500 g for 20 min at 20° with the brake off. The supernatant, the monoculear cells at the interface, and the layer above the pellet are carefully withdrawn. Care is taken to remove the mononuclear cells as completely as possible including those sticking to the wall of the tube. The cell pellet containing neutrophils and erythro-cytes is resuspended and washed in saline/low Ca^{2+} (300 g for 5 min at 4°).

The erythrocytes are then lysed with isotonic ammonium chloride. The cells are resuspended in 40 ml of an ice-cold solution of 155 mM NH_4Cl and 10 mM EDTA, pH 7.4. The tubes are kept on ice for 10 min with occasional mixing and are then centrifuged at 300 g for 5 min at 4°. The pellet is resuspended in saline/low Ca^{2+} and washed once (300 g for 5 min at 4°). If erythrocyte lysis appears incomplete, the treatment with ammonium chloride is repeated. The neutrophils are again washed and finally suspended in saline/low Ca^{2+} at a density of 10^8 cells/ml. The purity of neutrophils prepared in this manner is 90 % or higher.

B. Respiratory burst
1.00 pm

1. Oxygen uptake

The oxygen concentration of an isotonic medium (PBS) maintained at 37°C is determined by a polarographic method in an oxygraph cuvette. The Clark oxygen electrode is polarised at 0.6 V (Estabrook, 1967) and the signal corresponding to the oxygen level is continuously recorded. Under these conditions, the initial oxygen concentration is 205 µM. Neutrophils are added, and their O_2 uptake in the resting state is recorded. The cells are then stimulated, and the change in O_2 consumption is registered.

Place the following reagents into the incubation chamber :

First assay :
PBS/Ca^{2+}, Mg^{2+}	1.4 ml
Glucose	15 µl

Incubate for 5-10 min with stirring to allow temperature and oxygen equilibration, then add
Buffy coat 100 μl

Measure O_2 uptake until the rate becomes constant,
then add
KCN   10 μl

After 2-3 min, add
PMA   1 μl

Second assay :
PBS/Ca^{2+}, Mg^{2+} 1.4 ml
Glucose   15 μl
Equilibrate for 5-10 min, then add
Neutrophils (10^8 cells/ml) 100 μl, and sequentially :
PMA   1 μl
NaN$_3$ 10 μl
Cytochrome c   25 μl
SOD   10 μl

Record O_2 uptake for 2-3 min between additions.

Third assay :
PBS/Ca^{2+}, Mg^{2+} 1.4 ml
Glucose   15 μl
Equilibrate for 5-10 min, then add
Neutrophils (10^8 cells/ml)   100 μl

After 2-3 min add
fMLP 0.1 mM 15 μl

Record O_2 uptake until reaction has ceased, then add
PMA   1 μl

The aim of these experiments is to show that :
1. PMA induces a non-mitochondrial oxygen consumption.
2. O_2^- is produced.
3. The respiratory burst can be stimulated sequentially by stimuli with different specificity.

Usually, the PMA-induced oxygen uptake is 20-30 nmol/min per 10^7 neutrophils.

2. Superoxide production

Superoxide (O_2^-) production is determined by measuring superoxide dismutase (SOD) inhibitable cytochrome c reduction. Many electron donors besides O_2^- can reduce cytochrome c but only O_2^- reacts with SOD. The use of SOD confers specificity on this assay. Cytochrome c does not penetrate the neutrophil. For this reason only the O_2^- released into the extracellular medium will be detected.

2.1. Superoxide production by PMA-stimulated neutrophils

O_2^- production is followed continuously by measuring the change in absorbance of cytochrome c at 550 nm as a function of time. The reaction is performed in a double-beam spectrophotometer equipped with a thermostatted cell compartment.

The sample cuvette contains :

		Final concentration
PBS/Ca^{2+}, Mg^{2+}	2 ml	
Cytochrome c	20 µl	100 µM
Glucose	20 µl	10 mM
Catalase	5 µl	10 U/ml
Neutrophils	20 µl to 50 µl	10^6 cells/ml to 5×10^6 cells/ml

The reference cuvette contains :

PBS/Ca^{2+}, Mg^{2+} 2 ml
Cytochrome c   20 µl

Allow to equilibrate at 25°C in the thermostatted cell compartment (approximately 5 min). The reaction is started by adding 1 µl PMA to the sample cuvette, and the change in absorbance at 550 nm is recorded for about 5 min. Finally 10 µl SOD are added.

For calculating the rate of O_2^- production (nmol/min per 10^6 neutrophils) a reduced-minus-oxidized millimolar extinction at 550 nm of 21.0 for cytochrome c is used.

2.2. NADPH oxidase activity

The O_2^- forming oxidase is a membrane bound enzyme thought to be responsible for the respiratory burst of neutrophils. The enzyme catalyses the reaction :

$$NADPH \; + \; 2 \; O_2 \; \longrightarrow \; NADP^+ \; + \; 2 \; O_2^- \; + \; H^+$$

To demonstrate NADPH oxidase activity, the neutrophils are first activated with PMA and are then lysed with a detergent. O_2^- formation in the broken cells is induced by addition of NADPH. O_2^- formation is measured by the SOD inhibitable cytochrome c reduction.

The sample and reference cuvette are prepared as above (2.1.), except that 50 µl neutrophils (5 x 10^6 cells) are used. After equilibration at 25°C, 1 µl of PMA is added to the sample cuvette. Cytochrome c reduction is recorded until a constant rate is obtained. Next, 10 µl of deoxycholate are added to lyse the cells. O_2^- production stops, but after addition of 10 µl NADPH (or NADH) it is resumed. Finally, to demonstrate the SOD inhibitable portion of this reaction, 10 µl of SOD are added.

2.3. Comments

The NADPH oxidase of neutrophils is responsible for the sudden burst of oxygen consumption that follows exposure to a stimulating agent. The nature of the membrane associated enzyme is complex. The system is supposed to consist of at least two components, a flavoprotein and a cytochrome b with a low mid-point potential. The enzyme system catalyses the univalent reduction of molecular oxygen at the expense of pyridine nucleotide oxidation. NADPH or NADH can be used. However, NADPH is the preferred substrate as shown by the Km values of 30 μM for NADPH and 500 μM for NADH.

3. Hydrogen peroxide production

Superoxide produced during the respiratory burst is an instable intermediate that dismutates to form H_2O_2. Spontaneous dismutation is very rapid and is accelerated by SOD.

$$2\ O_2^{\bullet-} + 2\ H^+ \longrightarrow H_2O_2 + O_2$$

Several methods to measure H_2O_2 have been proposed. The most commonly used depends on the quenching of the fluorescence of scopoletin upon oxidation by horseradish peroxidase in the presence of H_2O_2. Despite high sensitivity, this method has the disadvantage that a decrease in fluorescence is measured which becomes non-linear as the scopoletin concentration decreases. The method that will be used here is based on the oxidation of p-hydroxyphenylacetate into a fluorescent dimer, which is mediated by horseradish peroxidase and dependent on H_2O_2. Other phenolic compounds may also be used as H-donors, e.g. homovanillic acid.

Place the following reagents into a fluorimetric cuvette :

First assay :
$$PBS/Ca^{2+}, Mg^{2+} \qquad 2 \text{ ml}$$
Glucose   20 µl
PHPA   35 µl
POD   15 µl

Mix, then add
Neutrophils (10^8/ml)   20 µl

Allow to equilibrate at 37°C in the cuvette holder for about 5 min. A standard amount of H_2O_2 (10 nmol) is added and the immediate increase in fluorescence (excitation wavelength 323 nm, emission wavelength 400 nm) recorded (calibration). The cells are then stimulated with

PMA   1 µl
Record H_2O_2 for 3-5 min, then add
SOD   10 µl

An increase in rate is observed. After 2-3 min add
Catalase   5 µl

Second assay :
Procedure as above except that 10 µl of SOD are added initially to the reaction mixture, and the cells are stimulated with 10 µl of fMLP instead of PMA.

C. Exocytosis

The Storage Compartments

Human neutrophils contain three types of storage organelles which discharge their content by exocytosis, the azurophil or primary granules, the specific or secondary

granules, and smaller secretory vesicles. Neutrophils from
other sources are similarly equipped. Azurophil and specific
granules have been identified in most species. Smaller storage
organelles have also been described, but so far secretory
vesicles were demonstrated only in human neutrophils. Ruminant
neutrophils appear to constitute a special case ; their major
storage organelle is a large, peroxidase-negative granule
distinct from the specific and azurophil granules. For all
neutrophils, the storage organelles are produced during the
period of maturation in the bone marrow. The mature cells
which are released into the circulation, and which usually are
the object of experimental studies, have lost the capacity to
form granules and are unable to replace the enzymes and other
proteins which they release.

Markers of Exocytosis

Azurophil granules contain a wide variety of acid
hydrolases, including N-acetyl-β-glucosaminidase, β-glucuroni-
dase, α-mannosidase, cathepsin B and D, and β-glycerophosphatase,
in addition to myeloperoxidase and two neutral proteinases,
elastase and cathepsin G. For a number of reasons, e.g. easy
assay conditions, stability and no tendency to adhere to
surfaces and cells, β-glucuronidase is the marker used most
commonly. Exocytosis from the specific granules can be best
assessed by assaying for vitamin B_{12}-binding protein. Lactofer-
rin, another exclusive constituent of the specific granules
has also been successfully used as a marker of release. It is
assayed by immunodiffusion or ELISA methods. Lysozyme is
present in both azurophil and specific granules and is
therefore not a useful marker for release from single
compartments of the neutrophil. Gelatinase is exclusively
contained in small storage organelles and is the marker of
choice for this compartment. Subcellular fractionation has
shown that the fractions containing these vesicles are
morphologically heterogeneous and also contain minor amounts

of acid hydrolases. Since the assay for gelatinase is compli-
cated and time consuming, release from this compartment will
not be considered here (see References).

1. Exocytosis experiment

Experimental procedure :

10^7 neutrophils are suspended in 0.9 ml PBS and are
preincubated either in the absence or presence of 5 µg/ml
cytochalasin B for 10 min at 37°C. They are then stimulated by
addition of 0.1 ml of the appropriate stimulus (see Table).
After further incubating for 15 min at 37°C the reaction is
stopped by rapid cooling in ice followed by centrifugation at
800 g for 10 min at 4°C. The supernatant is aspirated, diluted
with an equal volume of PBS containing 0.1 % Triton X-100, and
set aside for assays. A suspension of the pellet is prepared
with 1 ml of PBS containing 0.2 % Triton X-100, kept in ice
during 20 min with occasional mixing for extraction, diluted
with 3 ml of PBS and centrifuged at 800 g for 10 min at 4°.
The clear cell extract is used for assays. Similarly, an
extract from untreated cells is prepared for the determination
of the total cellular content of the markers.

Additions (ml)	Tube (N°)								
	1	2	3	4	5	6	7	8	9
PBS/Ca^{2+}, Mg^{2+}	0.8	0.8	0.8	0.8	0.8	0.7	0.7	0.7	0.7
Cytochalasin B (50 µg/ml)	-	-	-	-	-	0.1	0.1	0.1	0.1
Neutrophils (10^8/ml)	0.1	0.1	0.1	0.1	0.1	0.1	0.1	0.1	0.1
Preincubation 10 min at	4°	37°	37°	37°	37°	37°	37°	37°	37°
PBS	0.1	0.1	-	-	-	0.1	-	-	-
fMLP 1 µM	-	-	0.1	-	-	-	0.1	-	-
PMA 200 ng/ml	-	-	-	0.1	-	-	-	0.1	-
A 23187 10 µM	-	-	-	-	0.1	-	-	-	0.1
Incubation 15 min at	4°	37°	37°	37°	37°	37°	37°	37°	37°

Calculation of results :

In order to compare exocytosis from different subcellular compartments and under different conditions, release is
conveniently expressed as percentage of the total amount of
the respective marker. The validity of such figures depends
on reliable assay conditions and good recoveries. It is
therefore essential to calculate specific activities and
percentage recoveries in each exocytosis experiment.

Controls :

The accurate assessment of stimulus-dependent exocytosis requires comparison with exocytosis obtained under
control conditions, i.e., samples handled in the same way but
without addition of a stimulus and samples kept at 4°C.

Finally, it is important to test for the specificity
of the release observed. Exocytosis should occur without
damage to the cells, i.e. it should not be accompanied by a
loss of cytosolic components. The release of the cytosolic
enzyme lactate dehydrogenase is the most common way to test
for cell viability. It is important to recognize that a minor
amount of lactate dehydrogenase is always released under
control conditions.

2. Assay of markers

2.1. Vitamin B_{12}-binding protein (Specific granules)

A suspension of albumin-coated charcoal is prepared
by mixing equal volumes of a 1 % aqueous solution of BSA and
a 5 % suspension of Norit A charcoal in H_2O.

Assay :

50 µl sample, 200 µl 0.1 M potassium phosphate buffer, pH 7.5, and 500 µl of the diluted cyano [^{57}Co] cobalamin solution are mixed and incubated for 15 min in the dark at room temperature with occasional mixing. 1 ml of the albumin-coated charcoal suspension is then added, and after standing for another 15 min the mixtures are centrifuged at 2500 g for 15 min at 20°. The radioactivity of 1 ml of the supernatant is determined in a gamma-counter. The total activity of the diluted cyano [^{57}Co] cobalamin solution used is determined in a suitable aliquot. A sample blank is obtained by substituting PBS for the sample in the assay mixture and a total activity blank by omitting the cyano [^{57}Co] cobalamin solution. The amount of vitamin B_{12} bound by the sample is calculated by the formula

[0.05 x cpm (sample)]/cpm (total) = nanograms vitamin B_{12}

The respective blank values are deducted.

2.2. β-Glucuronidase (Azurophil granules)

Assay :

100 µl sample and 100 µl 4-methylumbellilferyl-β-D-glu-curonide are mixed and incubated for 15 min at 37°C. The reaction is stopped by addition of 3 ml stop solution. Liberated 4-methylumbelliferone is measured fluorimetrically (excitation : 365 nm, emission : 460 nm).

2.3. Lactate dehydrogenase (Cytosol)

Assay :

Into a 1 ml cuvette are added 0.5 ml of buffer, 0.2 ml of NADH, 0.1 ml Triton X-100, and 0.1 ml of sample. The reaction is started by addition of 0.1 ml of pyruvate and the

change in absorbance at 340 nm is recorded continuously.
Blanks are run in the absence of pyruvate.

IV. REFERENCES

Absolom D. (1986). Basic methods for the study of phagocytosis. In : **Methods Enzymol.**, **132** : 95-180.

Babior B.M. (1978). Oxygen-dependent microbial killing by
phagocytes. **New Engl. J. Med.**, **298** : 659-668, 721-725.

Baggiolini M. (1984). Phagocytes use oxygen to kill bacteria.
Experientia, 40 : 906-909.

Baggiolini M. and Dewald B. (1984). Exocytosis by neutrophils.
In : **Contemps. Top. Immunobiol.,** **14** : 221-246.

Dewald B., Bretz U. and Baggiolini M. (1982). Release of
gelatinase from a novel secretory compartment of human
neutrophils. **J. Clin. Invest.**, **70** : 518-525.

Dewald B. and Baggiolini M. (1986). Methods for assessing
exocytosis by neutrophil leukocytes. In : **Methods Enzymol.**,
132 : 267-277.

Doussiere J. and Vignais P.V. (1985). Purification and properties of an O_2^- - generating oxidase from bovine polymorphonuclear neutrophils. **Biochemistry, 24** : 7231-7239.

Estabrook W.R. (1967). Mitochondrial respiratory control and
the polarographic measurement of ADP : O ratios. In :
Methods Enzymol., **10** : 41-47.

Hyslop P.A. and Sklar L.A. (1984). A quantitative fluorimetric
assay for the determination of oxidant production of polymorphonuclear leukocytes : Its use in the simultaneous
fluorimetric assay of cellular activation processes. **Anal.
Biochem.**, **141** : 280-286.

Markert M., Andrews P.C. and Babior B.M. (1984). Measurement
of O_2^- production by human neutrophils. The preparation and
assay of NADPH oxidase-containing particles from human
neutrophils. In : **Methods Enzymol.**, **105** : 358-365.

Morel F. and Vignais P.V. (1984). Examination of the oxidase
   function of the b-type cytochrome in human polymorphonuclear
   leucocytes. **Biochim. Biophys. Acta, 764** : 213-225.
Rossi F. (1986). The O_2^- forming NADPH oxidase of the phagocytes : nature, mechanisms of activation and function.
   Biochim. Biophys. Acta, 853 : 65-89.

Experiment n° 4

ENDOCYTOSIS AND EXOCYTOSIS IN DICTYOSTELIUM AMOEBAE

G. KLEIN, M. BOF and M. SATRE

I. INTRODUCTION AND AIMS

The biological process of endocytosis is the ability
to internalize extracellular material in vesicles derived from
the plasma membrane and is a property shared by most
eucaryotic cells. The slime mold **Dictyostelium discoideum** is a
simple eucaryote which is quite well suited to study the
biochemistry and genetics of endocytosis. In nature, amoebae
feed exclusively by ingesting soil bacteria. For laboratory
use, axenic strains have been developed which entirely rely on
fluid-phase pinocytosis of nutritive medium to fulfill their
food requirements although they retain their entire phagocytic
potential toward bacteria, yeast, erythrocytes or other
particulate substrates. **Dictyostelium** amoebae grow rapidly and
homogeneous populations can be obtained easily in large
quantities. Cells are haploid, thus facilitating the isolation
of mutants. A parasexual system permits classical genetic
analysis and the molecular biological techniques can be
applied to **Dictyostelium** (Loomis, 1975 ; Raper, 1984 ; Nellen
et al., 1987).

The principal objectives of the described experiments
are to characterize kinetically fluid-phase endocytosis in
axenically grown **Dictyostelium** amoebae. A typical fluid-phase
marker, FITC-dextran, is utilized to determine precisely the
volume of fluid internalyzed via endocytosis. Uptake and
efflux rates of FITC-dextran are kinetically compatible with
the entry and concentration of the marker within an unique
endosomal compartment that exchanges with extracellular fluid.
Intracellularly accumulated FITC-dextran also provides quantitative informations regarding endosomal pH. Excretion of
lysosomal enzymes is another facet of endocytic machinery

which reflects both the rates of endosome-lysosome fusion and
of membrane recycling. This selective process is followed here
by the appearance in the extracellular space of acid phosphatase, a marker enzyme of primary lysosomes. A massive
extracellular release of the lysosomal content is obtained
after loading of the endosomal compartment with sucrose, a
non-metabolized sugar.

II. EQUIPMENT, CHEMICALS AND SOLUTIONS

A. Equipment

1. Access to

- a culture room (22 $\overset{+}{-}$ 1ºC) equipped with an orbital shaker
  (100-200 rpm)
- an Eppendorf centrifuge
- a refrigerated low-speed centrifuge equipped to spin 16 x
  50 ml or 36 x 15 ml
- a spectrophotometer
- a fluorometer
- a Coulter counter and C256 Channelizer
- a laminar flow hood.

2. On the bench

- orbital shaker (100-300 rpm)
- 5 and 10 µl Hamilton seringes
- sterile glassware (300 ml and 2 l) with metal caps
- sterile pipettes (1 and 10 ml)
- automatic pipettes (200, 1000 and 5000 µl)
- polyethylene screw-capped tubes (50 ml)
- disposable conical polystyrene tubes (15 ml)
- disposable plastic cuvettes (2 and 4 faces)
- disposable counting vials (Accuvettes, Coulter)
- 0.45 µ Millex-HA filters (Millipore SLHA 025 BS)

B. Microorganism

Dictyostelium discoideum, axenic strain AX-2 (ATCC 24397, obtainable from the American Type Culture Collection, 12301 Parklawn Drive, ROCKVILLE, MD 20852, USA).

C. Chemicals

- (^{3}H) dextran (Amersham)
- Fluorescein isothiocyanate (FITC)-dextran (FD 70, Sigma)
- Bovine serum albumin (Fraction V, Sigma)

D. Solutions

AX2 medium (Watts and Ashworth, 1970)

Oxoid peptone	14.3 g/l
Oxoid yeast extract	7.15 g/l
Na_2HPO_4, 2 H_2O	0.64 g/l
KH_2PO_4	0.48 g/l
dihydrosteptomycin	0.25 g/l
maltose	18.00 g/l
Sterilize 20 min, 120°C	

Counting medium (Isoton/5) :

Isoton II R (Coultronics)	400 ml
Formaldehyde 37 %	6 ml
H_2O	1 600 ml

Pi buffer, pH 6.3 :

Na_2HPO_4, 2 H_2O	0.36 g/l
KH_2PO_4	2.00 g/l

Wash media (ice cold) :
Pi buffer + 0.05 % BSA (w/v)
20 mM Mes-Na buffer, pH 6.3

Excretion media :

       20 mM Mes-Na buffer, 100 mM maltose, pH 6.3

       20 mM Mes-Na buffer, 100 mM sucrose, pH 6.3

Fluorescence buffers :

       50 mM Na_2HPO_4, 0.25 % Triton X-100 (w/w)

       50 mM NaPi buffers

       pH 4.5, 5.0, 5.5, 6.0, 6.5, 7.0, 7.5, 8.0, 8.5

Protein determination :

       SDS 10 % (w/v)

       Cu-EDTA reagent :

         A) cupric EDTA   0.5 g/l

         B) Na_2CO_3 40   g/l

         NaOH   8   g/l

       Just before use, mix equal volumes of (A) and (B)
Folin-Ciocalteu phenol reagent (BDH).

       1 mg/ml Bovine serum albumin (w/v)

III. EXPERIMENTAL PROCEDURES

1. Obtention of cells

1.1. Day -1

       Inoculate 4 x 300 ml-Erlenmeyer flasks containing 4 x
50 ml axenic medium AX2 at 22 $\pm$ 1°C with 4 x 7.5 ml of an
exponentially growing stock culture at about 7 x 10^6 cells/ml.

1.2. Day of the Experiment

Dilute a 50 µl-aliquot sample of culture in 10 ml of Isoton/5, count 0.5 ml of the diluted cells and record the size distribution with the Coulter counter and Channelizer. Take a total of 1 x 10^9 cells (about 200 ml of culture) and centrifuge at 4°C in 4 x 50 ml-polypropylene tubes, decant carefully supernatants and store cell pellets on ice. Tubes **(A)** and **(B)** are ready to start parts **(A)** and **(B)** of the experimental protocol. For tubes **(C)** and **(C')**, resuspend each pellet in 50 ml **ice cold** 20 mM Mes-Na buffer, pH 6.3 and centrifuge at 2400 rpm for 4 min. Decant the supernatants and repeat once the washing procedure. The cell pellets (tubes **(C)** and **(C')**) are stored on ice and ready for part **(C)** of the experimental protocol.

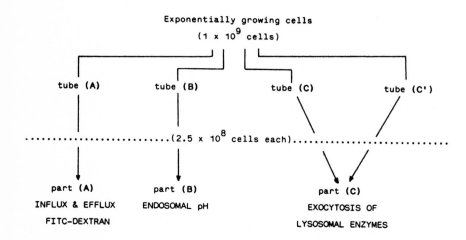

2. Part (A) : Kinetics of fluid-phase pinocytosis
   (Thilo and Vogel, 1980 ; Klein and Satre, 1986)

   2.1. Pinocytic influx : cell loading with FITC-
dextran

   Resuspend cell pellet from tube (A) in 50 ml of axenic
medium (AX2) at 22°C and transfer to a 500 ml-Erlenmeyer
flask. At t = 0, add 5 ml FITC-dextran (20 mg/ml) dissolved in
AX2 and filtered through a Millipore filter (0.45 μm). Agitate
on a shaker at 200 rpm. Take 1 ml-aliquots at t = 1, 10, 20,
30, 40, 60, 80, 100, 120, 150, 180 min (see time schedule) and
quench by dilution in 10 ml ice cold Pi-BSA.

   2.2. The pinocytic plateau is a dynamic equilibrium

   At t = 90 min, transfer 15 ml of the above FITC-dextran incubation medium to a 250 ml-Erlenmeyer flask containing
50 μCi (^{3}H) dextran (150 μl). Take 1 ml-aliquots at t = 91,
96, 101, 121, 141, 161, 181 min (see time schedule) and quench
with 10 ml ice cold Pi-BSA.

   2.3. Exocytosis of FITC-dextran

   At t = 103 min, take 20 ml of the above cells in
FITC-dextran medium, dilute to 50 ml with ice cold Pi-BSA,
centrifuge at 4°C, 2400 rpm for 4 min and wash with 50 ml ice
cold Pi-BSA. At t = 125 min, resuspend the cells in 20 ml AX2
medium at 22° C, transfer to a 250 ml-Erlenmeyer flask and
shake at 200 rpm. Take 1 ml-aliquots at t = 126, 130, 135,
145, 155, 165, 175, 185, 200 min (see time schedule) and
quench in 10 ml ice cold Pi-BSA.

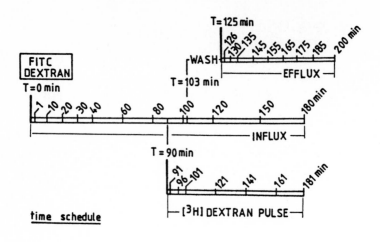

time schedule

2.4. Measurement of internalyzed material

At t = 200 min, centrifuge all the cells quenched in ice cold Pi-BSA and wash them twice with 10 ml of the same buffer by centrifugation at 2400 rpm for 4 min at 4°C. Cell pellets are suspended by vortexing in 1 ml of ice cold Pi buffer and 50 µl-aliquots of all the samples are diluted in 10 ml Isoton/5 and counted in the Coulter counter.

Add 2 ml of 50 mM Na_2HPO_4, 0.25 % Triton X-100 to the tubes from (2.1) (influx of FITC-dextran) and (2.3) (efflux). Measure the fluorescence of the samples in disposable plastic cuvettes (4 optical faces) with an excitation wavelength set at 470 nm and an emission wavelength set at 520 nm. Estimate internalyzed FITC-dextran by comparison with a calibration curve established with 0, 1, 2, 3, 4, 5, 6 µl of the incubation medium containing FITC-dextran + 1.9 ml Pi buffer + 4 ml 50 mM Na_2HPO_4, 0.25 % Triton X-100.

Transfer the content of the tubes from (2.2) to scintillation vials and add 10 ml of scintillation fluid. Count each vial for 2 min. Establish a calibration curve with 0, 1, 2, 3, 4, 5 µl of the incubation medium + 1 ml Pi buffer + 10 ml scintillation fluid. Estimate the amount of internalyzed (^{3}H) dextran from the calibration curve.

2.5. Calculations

Express results as an <u>Endocytic</u> <u>Index</u> : µl of the incubation medium pinocytosed per 10^6 cells vs time. Plot influx and efflux (as percentage of FITC-dextran remaining in the amoebae) and calculate the kinetic parameters : influx and efflux rates of FITC-dextran and influx rate of (^{3}H) dextran and the pinocytic capacity. Typical curves are shown in Fig. 1. Left panel : Influx of FITC-dextran (o) and (^{3}H) dextran (●). Right panel : Efflux of FITC-dextran.

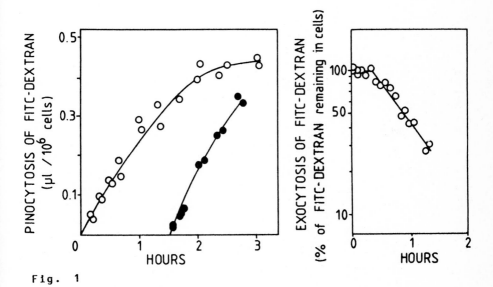

Fig. 1

3. Part (B) : determination of endosomal pH

The fluorescence intensity and the excitation spectrum of fluorescein or fluorescein conjugated to macromolecules are highly dependent on pH (Ohkuma and Poole, 1978 ; Heiple and Taylor, 1982). This can be exploited to determine the internal pH of the endosome in **Dictyostelium** amoebae since FITC-dextran is accumulated in this organelle during fluid-phase pinocytosis. The effects of a weak base on endosomal pH can be followed.

3.1. Protocol

In a first phase, load the endosomal compartment with FITC-dextran. Resuspend the cell pellet from tube **(B)** in 45 ml of AX2 medium at 22°C, transfer to a 500 ml-Erlenmeyer flask and shake at 200 rpm. At t = 0, start incubation by adding 5 ml FITC-dextran (20 mg/ml in AX2 medium) to **Dictyostelium** suspension.

Establish a calibration curve of FITC-dextran fluorescence as a function of pH. For this, in disposable plastic cuvettes (4 optical faces), add 5 μl of FITC-dextran (20 mg/ml) to 3 ml of 50 mM NaPi buffers at pHs 4.5, 5.0, 5.5, 6.0, 6.5, 7.0, 7.5, 8.0 or 8.5. Emission wavelength is set at 520 nm and fluorescence intensities are measured with excitation wavelength set at 450 nm or 490 nm. Plot the fluorescence intensities ratio (490 nm/450 nm) as a function of pH. The absolute fluorescence intensity ratios depend upon the instrument used. A typical sigmoidal curve obtained on a Kontron SFM 25 spectrofluorometer is shown in Fig. 2. The curve was fitted to the equation :

$$pH = 5.92 + \log \frac{(R - 0.854)}{(4.662 - R)}$$

where R is the fluorescence intensity ratio defined above.

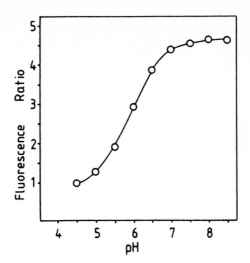

Fig. 2

For a quantitative estimation of the amount of internalyzed FITC-dextran, establish also a calibration curve of FITC-dextran fluorescence by adding 1, 2, 3, 4 or 5 µl of FITC-dextran containing incubation medium to 3 ml of 50 mM Na_2HPO_4, 0.25 % Triton X-100 and measure fluorescence intensity (emission wavelength = 520 nm ; excitation wavelength = 470 nm).

At t = 150 min, distribute 10ml-aliquots of cells to 5 x 15 ml conical polystyrene tubes, centrifuge and wash twice with **ice cold** Pi-BSA. Keep washed cell pellets **(3.1)** to **(3.5)** on ice until measurements.

Resuspend cell pellet from tube **(3.1)** in 3 ml 50 mM NaPi buffer pH 7.5. Dilute a 25 µl-aliquot in 10 ml Isoton/5 for cell counting with the Coulter counter. Measure **immediately** the fluorescence (to minimize exocytosis) with an emission wavelength of 520 nm and successively at the two excitation wavelengths of 490 nm and 450 nm. Add then 30 µl of 1 M imidazole-HCl, pH 6.0 and measure again the fluorescence after stabilization (after a couple of minutes).

Dilute a 300 µl-aliquot with 2.7 ml Na_2HPO_4, 0.25 %
Triton X-100 and measure fluorescence (emission wavelength =
520 nm ; excitation wavelength = 470 nm).

Repeat the same measurements with tubes (3.2) to
(3.5) and buffers at pH 7.0, 6.5, 6.0 and 5.5 respectively.

3.2. Calculations

Determine the amount of internalyzed FITC-dextran and
the endosomal pH in **Dictyostelium** amoebae from the pH
calibration curve and in the different experimental conditions.

4. Part (C) : excretion of lysosomal enzymes

A variety of eucaryotic cells are known to secrete
selectively lysosomal hydrolytic enzymes. This process is a
direct consequence of membrane recycling occuring during
endocytosis as the secreted hydrolases are those trapped in
the shuttling vesicles that return membrane from the secondary lysosome to the plasma membrane (Crean and Rossomando,
1979 ; Dimond et al., 1981).

4.1. Experimental protocol

At t = 0, resuspend the pellet from tube (C) in 25 ml
of 20 mM Mes-Na buffer, 100 mM maltose, pH 6.3 and the pellet
from tube (C') in 25 ml of 20 mM Mes-Na buffer, 100 mM
sucrose, pH 6.3 at 22°C in two 300 ml-Erlenmeyer flasks.
Shake at 200 rpm. For each condition and at t = 1, 10, 30,
60, 90, 120, 150 and 180 min, take two 1 ml-aliquots in two
Eppendorf tubes. Add 20 µl 10 % Triton X-100 to one of the
tubes and vortex immediately (Total = intra- and extracellular compartments). The other tube is centrifuged for 2 min in
an Eppendorf centrifuge. Transfer the supernatant to a new

Eppendorf tube containing 20 µl 10 % Triton X-100 (<u>Superna-</u>
<u>tant</u> : extracellular compartment). Store tubes on ice. For
each condition (maltose or sucrose) and for each time point,
measure both protein content and acid phosphatase activity in
the samples (<u>Total</u>, <u>Supernatant</u>).

4.2. Protein assay (Zak and Cohen, 1961)

Mix 0.1 ml aliquots of each sample with 0.2 ml of 10 %
SDS, 2.5 ml alcaline copper EDTA (solution A + B) and add
0.125 ml Folin-Ciocalteu reagent. Vortex immediately after
addition of the Folin reagent and incubate 30 min in the dark.
Measure optical densities at 660 nm. Protein amounts are
derived from a calibration curve established in the same
conditions with 10, 25, 50, 75 and 100 µg BSA.

4.3. Acid phosphatase assay (Wiener and Ashworth, 1970)

For each time point, transfer 50 µl of each sample
(<u>Total</u>, <u>Supernatant</u>) in maltose or sucrose conditions to
plastic disposable cuvettes. Include a blank containing 50 µl
H_2O with each series of samples. At 15 sec intervals, add 0.45
ml of assay medium (12 ml 0.5 M glycine-HCl buffer, pH 3.0 + 1
ml 0.24 M p-nitrophenylphosphate + 23 ml H_2O). Incubate for 20
min and stop the reaction by adding 2 ml 0.1 M NaOH. Measure
immediately the optical density in a spectrophotometer at 400
nm. Under these conditions, 1 µmol of p-nitrophenol formed
corresponds to an optical density at 400 nm of 7.52.

4.4. Calculations

For the maltose or sucrose media, plot the protein
data and the volumic activities of acid phosphatase in the
total suspension and in the extracellular compartments as
functions of time. Representative curves are shown in Fig. 3.

Panel A : total (o) and excreted (●) PNPase in maltose medium.
Total (Δ) and excreted (▲) PNPase in sucrose medium. Panel B :
same symbols for protein.

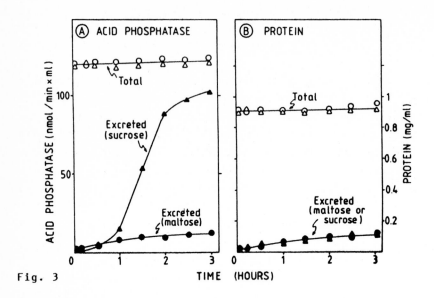

Fig. 3   TIME (HOURS)

IV. REFERENCES

Crean E.V. and Rossomando E.F. (1979). Effects of sugars on
    glycosidase secretion in **Dictyostelium discoideum**. J. Gen.
    Microbiol., **110** : 315-322.
Dimond R.L., Burns R.A. and Jordan K.B. (1981). Secretion of
    lysosomal enzymes in the cellular slime mold, **Dictyostelium
    discoideum**. J. Biol. Chem., **256** : 6565-6572.

Heiple J.M. and Taylor D.L. (1982). An optical technique for measurement of intracellular pH in single living cells. Intracellular pH : its measurement, regulation and utilization in cellular functions. (Nuccitelli N. and Deamer D.W., eds), pp. 21-54, Alan Liss, New-York.

Klein G. and Satre M. (1986). Kinetics of fluid-phase pinocytosis in **Dictyostelium discoideum** amoebae. **Biochem. Biophys. Res. Commun.**, **138** : 1146-1152.

Loomis W.F. (1975). **Dictyostelium discoideum.** A developmental system, pp. 1-214, Academic Press, New-York.

Nellen, W., Datta, S., Reymond, C., Sivertsen, A., Mann, S., Crowley, T. and Firtel, R.A. (1987). Molecular biology in Dictyostelium : Tools and applications. **Methods Cell Biol.**, **28** : 67-100.

Ohkuma S. and Poole B. (1978). Fluorescence probe measurement of the intralysosomal pH in living cells and the perturbation of pH by various agents. **Proc. Natl. Acad. Sci. USA** **75** : 3327-3331.

Raper K.B. (1984). The Dictyostelids, pp. 1-453, Princeton University Press.

Thilo L. and Vogel G. (1980). Kinetics of membrane internalization and recycling during pinocytosis in **Dictyostelium discoideum. Proc. Natl. Acad. Sci. USA 77** : 1015-1019.

Watts, D.J. and Ashworth, J.M. (1970). Growth of myxamoebae of the cellular slime mould Dictyostelium discoideum in axenic culture. **Biochem. J.**, **119** : 171-174.

Wiener E. and Ashworth J.M. (1970). The isolation and characterization of lysosomal particles from myxamoebae of the cellular slime mold **Dictyostelium discoideum. Biochem. J.**, **118** : 505-512.

Zak B. and Cohen J. (1961). Automatic analysis of tissue culture proteins with stable Folin reagents. **Clin. Chim. Acta, 6** : 665-670.

Experiment n° 5

ISOLATION OF INTACT CHLOROPLASTS - CRITERIA OF INTEGRITY

R. DOUCE, J. JOYARD and M. NEUBURGER

I. INTRODUCTION AND AIMS

Intact chloroplasts are defined as those with an intact envelope as revealed by the ability to carry out light-dependent CO_2 fixation ; ribonuclease and protease resistant RNA and protein synthesis ; starch, fatty acids and amino acids synthesis ; and to show low rates of ferricyanide reduction which are greatly increased by exposure to hypotonic conditions (for a review see Douce and Joyard, 1979).

In leaf cells, the active cytoplasm occupies a peripheral shell between the central vacuole and a rigid cell wall (Gunning and Steer, 1975). The vacuole is a large watery compartment surrounded by a thick membrane (tonoplast) and containing a wide variety of substances harmful to chloroplast (flavonoids, various colorless phenolic compounds) and hydro-lases (lipases, proteinases, etc.). When leaves are homogeni-zed in order to isolate chloroplasts, cellular compartmentali-zation is destroyed and the "secondary products" are released into the medium with effects ranging from undesirable to devasting depending on the species and on the isolation technique. Enzymes that liberate free fatty acids such as linoleic and linolenic acids, the two major fatty acids of plant cell membranes, are potentially the most troublesome contaminants in chloroplast preparation from leaves. In most case, these fatty acids are produced by the actions of lipolytic acyl hydrolases released during the course of the tissue grinding. These enzymes cause a very rapid hydrolysis of chloroplast lipids (galactolipids, sulpholipid, phospholi-pids) leading to functional impairment. In addition, unsatura-ted fatty acids thus released are the best natural substrates for lipoxygenase (an enzyme that catalyzes the addition of O_2 to linoleic acid), producing harmful fatty acid hydroperoxides. Apart from problems associated with the presence of inhibiting substances, there is a need for strong shearing forces to disrupt the rigid cell wall. These forces are also detrimental to the chloroplasts. Finally it is strongly recommended that investigators use mature leaves grown under controlled environmental conditions in order to reduce the number of large starch granules. The plant material should therefore not be intensely illuminated for long period prior to chloroplast isolation. Chloroplasts containing large starch grains will generally be broken during centrifugation.

Consequently, expertise is required in order to prepare chloroplasts displaying the same biochemical and morphological characteristics as chloroplasts **in situ** (Walker, 1987). We shall first describe a procedure for the isolation of intact chloroplasts which is currently in use and then define the criteria for the assessment of chloroplast integrity including (a) the light-CO_2-dependent O_2 evolution and (b) the permeability of the chloroplast envelope to ferricyanide.

II. EQUIPMENT, CHEMICALS AND SOLUTIONS

A. Equipment

1. Access to

- a cold room (+ 4°C)
- a superspeed centrifuge, refrigerated (RC 5, Sorvall or equivalent) with the Sorvall ARC-1 automatic rate controller and the following rotors :
  - fixed angle rotor, 8 x 50 ml (SS 34, Sorvall or equivalent)
  - vertical rotor, 8 x 36 ml (SV-288 formerly SS-90, Sorvall or equivalent)

- a spectrophotometer allowing measurements at 750 nm
- a high speed homogenizer with a 200 ml container (e.g.,
  Polytron Kinematica, type 35/2M probe, CH 6010 Kriens/
  Luzern)
- a 150-w Xenon arc lamp source (Oriel Corporation)

2. On the bench

- a magnetic stirrer with Teflon-coated magnetic rods
- an oxygen electrode (Hansatech limited, Paxman Road Hardwick
  Industrial Estate, Kings Lynn, Norfolk)
- Glassware, Hamilton syringues (5 to 50 μl)
- Nylon blutex, 50 μm pore size (Tripette et Renaud)
- Muslin or Cheesecloth, 60 cm large (Ruby)
- Scissors
- Corex tubes, 30 ml for centrifugation

B. Chemicals

- 6-8 week-old deribbed spinach leaves and fully expanded pea
  leaves were supplied by the Botanical station (Besançon)
- D(-)-sorbitol for microbiology (references 7758, Merck)
- Hepes or 4-(2-hydroxyethyl)-1-piperazin-ethane sulfonic acid
  (Reference 737-151, Boehringer)
- EDTA or ethylenediamine tetraacetic acid, free acid,
  purified grade (reference ED, Sigma)
- Tetra-sodium pyrophosphate, for analysis (reference 6591,
  Merck)
- Albumine bovine, fraction V (reference A-8022, Sigma)
  previously defatted (this was achieved by successive washing
  of 100 g albumine by 1 liter of ethanol and 1 liter of
  acetone, the albumine is then recovered by filtration on a
  buchner funnel)
- Percoll TM (Pharmacia Fine Chemicals)
- NaOH, HCl, $MgCl_2$ and $MnCl_2$, for analysis

C. Solutions

The following solutions should be prepared in advance, kept at 0-4°C.

- Isolation medium :

Sorbitol (330 mM), EDTA (2 mM), Tetra-sodium pyrophosphate (30 mM adjust to pH 7.8 with HCl), 0.1 % defatted albumine bovine.

- Washing medium :

Sorbitol (330 mM), Hepes (50 mM, adjust to pH 7.8 with KOH), EDTA (2 mM), 0.1 % defatted albumine bovine.

- Percoll medium :

50 % (v/v) Percoll, Sorbitol (330 mM), Hepes (50 mM, adjust to pH 7.8 with KOH), EDTA (2 mM), 0.1 % defatted albumine bovine.

- O_2-electrode medium :

Sorbitol (330 mM), EDTA (2 mM), $MgCl_2$ (1 mM), $MnCl_2$ (1 mM), Hepes (50 mM, adjust to pH 7.8 with KOH). It is also necessary to prepare a double-strength medium (see below).

- Potassium ferricyanide (50 mM)
- K-phosphate (200 mM ; 10 mM)
- $NaHCO_3$ (500 mM)
- NH_4Cl (250 mM)

III. EXPERIMENTAL PROCEDURES

If the investigations to be undertaken do not demand a particular species, difficulties can usually be minimized by careful choice of plant material (spinach leaves, pea leaves, etc.).

A. Preparation of spinach leaf chloroplasts
9.00 am ──────→ 12.00

In each experiment, 50 g of 6-8 week-old deribbed spinach leaves are chopped with scissors and ground with 250-300 ml isolation medium for 1-5 seconds in a perspex vessel using a polytron homogeniser (Kinematica, Switzerland). Longer blending improves the yield of recovered chloroplasts but considerably increases the percentage of envelope-free chloroplasts and chloroplasts that have resealed following rupture and the loss of stroma content. The aim is therefore to break as many cells but as few chloroplasts as possible. All operations are carried out in the cold (0-2°C) by keeping samples on ice. Furthermore, it is desirable to have the isolation medium as a semi-frozen slush to reduce cavitation.

The brei is rapidly filtered through 6 layers of muslin and one layer of nylon blutex (50 micron pore size). The chloroplasts should then be separated from the filtrate (approx 200-250 ml) as quickly as possible. The filtered suspension is centrifuged (8 tubes, 30 ml each) for 3 min at 1 500 x g before being braked as quickly as possible (if the tubes are only half-filled the centrifugal path is reduced and this improves the yield of intact chloroplasts). The supernatant is poured off and each pellet is resuspended, using either a small spatula or a rod wrapped with cotton wool by addition of small quantity (0.5 ml) of washing medium. The suspension (total volume 4-5 ml) is filtered on a nylon blutex in order to separate the aggregates. The nylon is then washed with a minimal volume (approximately 1 ml) of washing medium.

The chloroplasts are normally resuspended to give a chlorophyll concentration of 2-4 mg per ml and stored on ice.

At this stage a careful examination by phase contrast
light microscopy of the crude chloroplast pellet obtained by
differential centrifugation shows that the procedure described
above results in a preparation which contains 50-90 % intact
chloroplasts, depending on the skill of the experimenter, but
which is also contaminated to some extent, by various cell
organelles. Intact chloroplasts are refractile with a bright
halo apparent around each plastid whereas broken chloroplasts
are dark-green and non-refractile. Purification of intact
chloroplasts is therefore necessary.

B. Purification of intact chloroplast

The recent introduction of a non-toxic silica sol
gradient material Percoll TM (colloïdal silica coated with
polyvinylpyrrolidone) by Pharmacia has permitted the development of a rapid purification procedure utilizing iso-osmotic
and low viscosity conditions. The following method, based on
that of Mourioux and Douce (1981) yields preparations of
chloroplasts from peas or spinach which are greater than 95 %
intact and capable of high rates of CO_2 fixation.

Thirty-six ml of the 50 % percoll medium are pipetted
into 8 centrifuge tubes. The tubes are placed in a precooled
Sorvall SS-90 rotor and centrifuged (at 4°C) at 10,500 rpm
(10,000 g) for 100 min the rate controller is essential in the
last steps of deceleration. At the conclusion of this step a
continuous gradient of Percoll is obtained in each tube. The
upper 2 ml of the gradient are removed and aliquots (2 ml
sample, 4 to 6 mg chl) of the crude chloroplast suspension
then are layered on the linear Percoll gradients. The tubes
are centrifuged (the automatic rate controller must be used)
for 10 min at 5,000 x g (at 4°C) in the Sorvall SV-288 rotor.

Intact chloroplasts (class A) are recovered as a broad band
near the bottom of the tube whereas fragmented chloroplasts
(class C or stripped chloroplasts), other cell debris and
organelles formed a band at the sample-gradient interphase and
large contaminants recovered as dark line along the side of
the tube.

In all the cases, the stripped chloroplasts are
removed by aspiration and the intact chloroplasts collected
carefully ; for instance, it is essential not to resuspend the
pellet which sticks on the side of the tube. The intact
purified chloroplasts are diluted with washing medium (10
volumes to 1 volume of chloroplast suspension) and recovered
as a pellet after centrifugation (3500 x g for 90 s, at 4°C,
SS 34 rotor, Sorvall). The supernatant is removed by aspiration and the pellet of purified chloroplasts is gently
resuspended in few drops of washing medium (you may also find
some advantage in storing your chloroplasts as a pellet until
immediately prior to use - a procedure which may diminish loss
of essential ions, such as inorganic phosphate, to the medium
as shown by Mourioux and Douce, 1981).

C. Criteria for the assessment of chloroplast integrity

Ferricyanide-dependent O_2 evolution

Although ferricyanide is a widely used Hill oxidant it
does not cross the intact chloroplast envelope and its
reduction is therefore only catalysed by the thylakoid

membranes present in the preparation and not by intact chloroplasts :

$$2 H_2O \rightarrow O_2 + 4 H^+ + 4 e^-$$
$$4 Fe(CN)_6^{3-} + 4e^- \rightarrow 4 Fe(CN)_6^{4-}$$

$$2 H_2O + 4 Fe(CN)_6^{3-} \rightarrow O_2 + 4 H^+ + 4 Fe(CN)_6^{4-}$$

Consequently, the intactness of chloroplast preparations can be rapidly evaluated by following ferricyanide-dependent O_2 evolution in the presence of 2 mM NH_4Cl (a potent uncoupler) before and after an osmotic shock (Heber and Santarius, 1970 ; Walker, 1987). The light is provided by a 150-w Xenon arc lamp source giving an irradiance of 700 w m^{-2} at the surface of the vessel. The measurement of oxygen evolution in a closed system (1 to 3 ml) is based on a device designed by Delieu and Walker (1981) and manufactured by Hansateck (O_2 electrode)[1].

---

[1] Normal calibration procedure :

- with O_2-electrode medium in the vessel set an arbitrary "air-line" by adjusting the electrical output of the electrode until the recorder pen is near the top of the chart.
- flush empty vessel with Argon gas to establish the "Argon-line".
- add O_2-electrode medium saturated with air at the operating temperature (25°C) and adjust a true "air-line" electrically so that it is a convenient multiple of the number of µmoles of O_2 per ml of electrode medium (at 25°C there are 0.25 µmole O_2 per ml).

The O_2-electrode medium is made double strength i.e. 660 mM sorbitol, 4 mM EDTA, 2 mM $MgCl_2$, 2 mM $MnCl_2$, 100 mM Hepes-KOH, pH 7.8. For intact chloroplasts add 0.5 ml double strength medium, 20 μl of 50 mM potassium ferricyanide, 50 μl of 200 mM Na-phosphate (in order to inhibit CO_2-dependent O_2-evolution), 0.5 ml water to give a total volume of 1 ml and lastly the chloroplasts (20-60 μg chlorophyll). The medium is bubbled with argon to remove O_2. After illumination (700 w.m^{-2}) for 30 s - 1 min, add 20 μl of 250 mM NH_4Cl to uncouple electron transport. For broken chloroplasts the procedure is repeated but the order of addition altered to measure total activity in osmotically shocked chloroplasts. Intact chloroplasts were added to 0.5 ml water (to give at least a 10-fold dilution) and stirred for 5 s before adding 0.5 ml double-strength medium to restore isotonic conditions following the osmotic shock and the other reagents. Under these conditions, the chloroplast envelope bursts and exogenous ferricyanide has access to the outer surface of the thylakoids where it is reduced triggering full O_2 evolution (1 μmole of O_2 is evolved for every 4 μmoles of ferricyanide added). Therefore it should be possible to estimate very rapidly the proportion of intact chloroplast in a given preparation from the relative rate of ferricyanide-dependent O_2 evolution by untreated and by osmotically shocked chloroplasts : if the intact chloroplasts give a rate of 10 and after osmotic shock the rate is 250, the percent intact is 100 (250-10/250) = 96 %.

CO_2-dependent O_2 evolution

It is now well established that ruptured chloroplasts (class C) are normally capable of O_2 evolution and photophosphorylation in the presence of appropriate oxidants such as ferricyanide (see above) and cofactors but, in contrast with intact chloroplasts, are inactive in CO_2-dependent O_2 evolution or CO_2-fixation (Walker, 1976). In fact, intact chloroplasts (self-sufficient photosynthesizing organelles) contain a large amount of soluble enzymes in the stromal compartment (about 9.5 g protein per g of chlorophyll) involved in CO_2 fixation which are rapidly released into the medium when the envelope membranes are damaged. Consequently the best indication of chloroplast integrity is to measure CO_2-dependent oxygen evolution rate and compare this with that of the parent tissue (150-200 µmol/h per mg chlorophyll).

Add 1 ml O_2-electrode medium, 10 µl of 500 mM $NaHCO_3$, 10 µl of 10 mM Na_2HPO_4, and chloroplasts equivalent to 70 to 80 µg Chl. The reaction medium is gassed with argon before the addition of chloroplasts and equilibrated with the O_2 electrode at 25°C. The light is provided by a 150-w Xenon arc lamp source giving an irradiance of 1,300 $w.m^{-2}$ at the surface of the vessel. When the mixture is illuminated, CO_2-dependent O_2 evolution normally starts after a lag phase (which largely reflects autocatalytic build-up of Benson-Calvin cycle intermediates). If small quantities of Pi are used, it will be seen that there is a stoichiometric relationship between added Pi and extent of the restoration in rate. This is consistent with the formation of triose phosphate according to the following equation[2] :

$$3\ CO_2 + H_2O + Pi \longrightarrow triose\ phosphate + 3\ O_2$$

---

[2] triose phosphate is the principal product of the photosynthesizing chloroplasts and there is an obligate exchange catalyzed by the phosphate translocator, between external Pi and internal triose phosphate (Walker, 1976 ; Heldt, 1976).

Conversely, if there is too much Pi in the external medium, triose phosphate which would have otherwise been used to regerenate the CO_2 acceptor (Ribulose 1,5-bisphosphate), is "pulled-out" of the chloroplast via the phosphate translocator and the lag phase is prolonged indefinitely.

D. Chlorophyll determination

Chlorophyll is determined by diluting a small aliquot of chloroplasts (10 µl) with 10 ml 80 % acetone in a corex centrifuge tube followed by centrifugation (10 min, 5 000 g) to remove precipitated proteins. Using the extinction coefficient of Bruinsma (1961), absorbance at 652 nm (1 cm light path) multiplied by the dilution factor (1000) and divided by 36 gives the chlorophyll concentration of the original chloroplast suspension in mg per ml.

IV. CONCLUSION

The method described above has been used with great efficiency for numerous C_3 plant species. However, mechanical procedures (e.g. involving the grinding of leaf tissues) have not been very successful for chloroplasts from C_4 or CAM plants. Therefore, numerous authors have used protoplasts prepared from leaf tissue by enzymatic digestion to obtain functional mesophyll chloroplasts form C_4 plants (see for instance Edwards and Walker, 1983). However, the apparent failure of the mechanical procedures for C_4 plants is likely to be due to unproper medium used rather than the fragility of the chloroplasts : for instance, Jenkins and Russ (1984), by using media containing excess amounts of EDTA were able to prepare mesophyll chloroplasts from several C_4 plants showing 80-90 % intactness and high rates of substrate dependent oxygen evolution under illumination. Therefore, one should keep in mind that modification of the composition of the grinding or washing medium are often a convenient way to improve the physiological integrity of chloroplasts during their isolation.

V. REFERENCES

Bruinsma J. (1961). A comment on the spectrophotometric determination of chlorophyll. **Biochim. Biophys. Acta,** 52 : 576-578.

Douce R. & Joyard J. (1979). Structure and function of the plastid envelope. **Adv. Bot. Res.,** 7 : 1-116.

Edwards G. & Walker D.A. (1983). C_3, C_4 : mechanical, and cellular and environmental regulation, of photosynthesis. **Blackwell Sci.** Publ. Oxford.

Gunning B.E.S. & Steer M.W. (1975). Ultrastructure and the Biology of Plant Cells. Arnold, London.

Heber U. & Santarius K.A. (1970). Direct and indirect transfer of ATP and ADP across the chloroplast envelope. **Z. Naturforsch,** 25 : 718-728.

Heldt H.W. (1976). Metabolite transport in intact spinach chloroplasts. in **"The intact Chloroplast"** (Barber J. ed) pp. 215-234, Elsevier, Amsterdam.

Jenkins C.L.D., & Russ V.J. (1984). Large scale, rapid preparation of functional nesophyll chloroplasts from zea mays and other C_4 species. **Plant Sci. Lett.,** 35 : 19-24.

Mourioux G. & Douce R. (1981). Slow passive diffusion of orthophosphate between intact isolated chloroplasts and suspending medium. **Plant Physiol.,** 67 : 470-473.

Walker D.A. (1976). CO_2 fixation by intact chloroplasts : Photosynthetic induction and its relation to transport phenomena and control mechanisms. in **"The Intact Chloroplast"** (Barber J. ed) pp. 235-278, Elsevier, Amsterdam.

Walker D.A. (1987). The use of the oxygen electrode and fluorescence probes in simple measurements of photosynthesis. **Research institute for photosynthesis,** University of Sheffield (distributed by Packard Publishing Limited, 16, Lynch Down, Funtington, Chichester, West Sussex PO18 9LR).

Experiment n° 6

METHODS FOR THE ISOLATION OF HIGHER PLANT VACUOLES

A. PUGIN, A. KURKDJIAN, D. LE QUÔC, K. LE QUÔC

I. INTRODUCTION AND AIMS

The central vacuole of the plant cell is a large
membrane-bound organelle which occupies up to 90 % of the
total cell volume. There is evidence indicating that the
vacuole participates in preservation of cytoplasmic homeosta-
sis, cell turgor, accumulation of ions and metabolites
(sugars, aminoacids, organic acids) ; secondary metabolites
such as alkaloids are also stored in this compartment (for
reviews, see Matile, 1978 ; Leigh, 1983 ; Boller and Wiemken,
1986).

The biochemical characterization of plant vacuoles
began in 1968 when Matile and Pujarniscle respectively
reported the isolation of small vacuoles from root tips of
maize (Matile, 1968) and of lutoids from the latex of **Hevea
brasiliensis** (Pujarniscle, 1968). The lutoids are intermediate
between lysosomes and plant vacuoles. Interest in the composition and function of plant vacuoles increased with the
publication in 1975-1976 of two different methods for isolating large quantities of mature plant vacuoles. These organelles were obtained from protoplasts (Wagner and Siegelman,
1975) or directly from storage root tissue (Leigh and Branton,
1976).

The basic method presented by Wagner and Siegelman
(1975) involves osmotic lysis of protoplasts. Vacuoles can be
prepared by this method from any tissue that yields protoplasts. The experimental procedure and the variations have
been described in details and commented by Wagner (1983).

Different methods have been used to prepare vacuoles from protoplasts under isotonic conditions : the rupture of the plasmalemma has been obtained by physical processes (Lörz et al., 1976 ; Martinoia et al., 1981) or chemical treatments : DEAE dextran (Buser and Matile, 1977 ; Schmidt and Poole, 1980 ; Boudet et al., 1981 ; Alibert et al., 1982), digitonin (Le Quoc et al., 1987).

The method for isolating vacuoles directly from tissue described by Leigh and Branton (1976) appears to be restricted to storage root tissue. It consists of mechanical slicing of fresh tissue (Leigh and Branton, 1976, Marty and Branton, 1980) or of plasmolysed tissue (Doll et al., 1979 ; Grob and Matile, 1979) using a slicing device. Released vacuoles are collected in a suitable medium and then purified by flotation or sedimentation in a discontinuous density gradient.

The preparations of vacuoles must be assayed for their purity. Unbroken protoplasts can be observed by photonic microscopy. Cytoplasmic contaminants and plasmalemma can be detected by fluorescence microscopic observations using in the first case fluorescein diacetate and in the second case concanavalin A fluorescein labeled as markers (Admon et al., 1980, Boller et al., 1976). The activity of marker enzymes of other cell compartments allows to assess the purity of isolated vacuoles. An extensive review of various marker enzymes for monitoring specific subcellular fractions of plants has been presented by Quail (1979).

We present three methods for the preparation of vacuoles from protoplasts. These methods mainly differ in the process used to lyse the plasmalemma. In the first method, the ultracentrifugation of protoplasts on density gradient allows in one step the liberation and the separation of two populations of vacuoles (Pugin et al. 1986). In the second method the lysis of the plasmalemma is obtained by a chemical treatment using digitonin, the vacuoles are separated by flotation (Le Quoc et al., 1987). In the third method the vacuoles are liberated by osmotic shock and separated by flotation. Fluorescence microscopy and assays with marker enzymes show that these vacuole preparations contain only traces of contamination by other organelles and by cytoplasmic components.

METHOD 1 : ISOLATION OF VACUOLES BY ULTRACENTRIFUGATION OF PROTOPLASTS ON DENSITY GRADIENT

II. EQUIPMENT, BIOLOGICAL MATERIAL, CHEMICALS AND SOLUTIONS

A. Equipment

1. Access to

- a balance
- an ice machine
- a horizontal thermostated waterbath shaker
- a refrigerated centrifuge (up to 3000 g) with a rotor of 200 ml capacity (e.g., Sigma 3 K2)
- a refrigerated ultracentrifuge (up to 200 000 g) with a rotor of 200 ml capacity (e.g., Beckman L8 70, rotor 70 TI)
- a microscope equipped for epifluorescence : filter I_2 exc. 490 nm lecture 525 nm (e.g., Reichert Jung Microstar)
- a spectrophotometer UV-vis
- an oxygraph

2. On the bench

- a glass filter (40/90 µm)
- a petri dish
- a 30 µm mesh nylon cloth
- 2 centrifuge tubes (95 x 28 mm)
- 6 conical centrifuge tubes (80 x 26 mm)
- 8 centrifuge bottles (92 x 25,4 mm)
- Pasteur pipettes and graduated glass pipettes
- a plastic syringe (10 ml)
- automatic pipettes (5 to 200 µl)
- a Nageotte haematocytometer
- a Malassez haematocytometer
- glass cuvettes for spectroscopy

B. Biological material

Acer pseudoplatanus isolated cells : the strain comes from a cambial tissue culture obtained by Lamport in 1964. Cells are cultivated in agitated liquid medium. The maintenance of the culture is ensured by the transfer of cells to fresh medium every 8 days.

C. Chemicals

Mannitol ; 2(N-morpholino)ethanesulfonic acid (MES) ; tris (hydroxymethyl) aminomethane (Tris) ; ethylenediamine-tetraacetic acid (EDTA) ; Fluorescein diacetate (FDA) ; Concanavalin A fluorescein labeled (all from Sigma), Ficoll 400 (Pharmacia), Macerozyme R 10 (Yakult Honsha Co., LTD, Tokyo), Cellulase Y-C, Pectolyase Y-23 (Sheishin Pharmaceutical Co., LTD, Tokyo).

D. Solutions

The following solutions should be prepared in advance and kept at 4°C.

1 - Enzyme solution for the preparation of protoplasts : 2 % Cellulase Y-C, 1.5 % Macerozyme R 10, 0.25 % Pectolyase Y-23, 0.6 M Mannitol dissolved in water at pH 7.5 and then adjust to pH 5.5 with 25 mM MES and 1 N HCl.
2 - Mannitol 0.7 M pH 6.0
3 - Buffer 1 : 12.5 mM MES, 12.5 mM Tris, 0.7 Mannitol, 1 mM EDTA, pH 7.0
4 - Buffer 2 : buffer 1 without EDTA
5 - Ficoll 12 %, Ficoll 4 % in buffer 1
6 - Fluorescein diacetate : 0.5 % in acetone ; final concentration in protoplast and vacuole preparations : 0.01 %

III. EXPERIMENTAL PROCEDURE

The description of the method for the isolation of vacuoles by ultracentrifugation was published by Pugin et al. (1986).

1. Preparation of protoplasts :

Cells (ca 5 g fresh weight) are harvested on a sintered glass filter (40/90 µm) under partial vacuum and rinsed with 20 ml 0.7 M Mannitol pH 6.0. The cells are transfered in a petri dish with 10 ml of enzyme solution containing 2 % Cellulase, 1.5 % Macerozyme, 0.25 % Pectolyase, 0.6 M Mannitol, 25 mM MES (pH 5.5) and shaked (40 strokes min^{-1}) for 1 hour at 30°C.

8.45 am The liberated protoplasts are filtered through a 30 µm mesh nylon cloth, washed twice with 40 ml buffer 2 and then collected by centrifugation (1500 g ; 4°C ; 1 min ; Sigma 3

K_2). The pellet of protoplasts is gently resuspended to obtain about 20.10^6 protoplasts ml^{-1}. The density of the population is measured using a Nageotte haematocytometer.

9.30 am 2. Preparation of vacuoles

The protoplasts are submitted to an ultracentrifugation in a discontinuous gradient of Ficoll 400 diluted in buffer 1. In centrifuge bottles (92 x 25,4 mm) are successively layered : 6 ml of 12 % (w/v) Ficoll, 6 ml of 6 % Ficoll containing 12.10^6 protoplasts, 8 ml of 4 % Ficoll and 5 ml of buffer 1. The tubes are centrifuged for 1 h at 200 000 g (Beckman L8 70 ; rotor 70 TI ; 4°C). The liberated vacuoles floated to the interphases between 6 % Ficoll and 4 % Ficoll (interphase 6/4) and between 4 % Ficoll and buffer (interphase 4/0). The vacuoles are harvested using a Pasteur pipette, diluted 3 times with buffer 2 and then collected by centrifugation (300 g ; 6 min ; 4°C Sigma 3 K_2) in conical centrifuge tubes (80 x 26 mm). The pellet of vacuoles is gently resuspended in buffer 2. The density of the populations of vacuoles is measured using a Malassez haematocytometer.

11.45 am 3. Fluorescence microscopy

Fluorescein diacetate is prepared as a 0.5 % solution in acetone and added to the preparations of protoplasts or vacuoles at a final concentration of 0.01 %. Preparations are observed 5 min later using a microscope (Reichert Jung Microstar) equipped for epifluorescence (filter I_2, exc. 490 nm, lecture 525 nm).

Concanavalin A fluorescein labeled is added to the preparations of protoplasts or vacuoles (5 µg per 10^5 organelles). Five minutes later the suspensions are centrifuged (1000 g ; 6 min ; 4°C ; Sigma 3 K_2). The supernatants are discarded, the collected organelles are resuspended in buffer 2 and observed as described above.

2.30 pm 4. Marker enzymes

Catalase, glucose-6-phosphate dehydrogenase and cytochrome c oxidase activities are measured according to standard technics ; α-mannosidase is measured according to Boller and Kende (1979). This enzymatic activity is used to measure both the yield of the method for the isolation of vacuoles and the number of vacuoles per protoplast. This number represented the ratio : α-mannosidase activity per 10^6 protoplasts/α-mannosidase activity per 10^6 vacuoles. UDP-glucose sterol glucosyltransferase (UDPG-ST) and Antimycin A insensitive NADH cytochrome c reductase activities are measured in the membranous fraction of protoplasts and in the preparation of whole vacuoles ; the first according to a method described by Hartmann-Bouillon and Benveniste (1978) and modified by Chanson et al. (1984) with 17,5 nCi UDP-[^{14}C] glucose per assay ; the second according to Schumaker and Sze (1985).

IV. COMMENTS

The ultracentrifugation on density gradient of protoplasts allows in one step the liberation and separation of two populations of vacuoles with a high yield (more than 50 %). The two populations of vacuoles are different because of their density and their size. The "light" vacuoles (interphase 0/4) are 16.50 μm $\pm$ 1.00 across the "heavy" vacuoles (interphase 4/6) are 10.70 $\pm$ 1.00 across. The number of vacuoles in each population changes with the state of the cell growth : with the same number of protoplasts coming from cells taken out 6, 7, 8 and 9 days after the beginning of the culture, the number of large vacuoles increased as the cells become older, whereas the quantity of the small vacuoles decreased. This separation of two populations of vacuoles can overcome the heterogeneity of cell suspensions provoked by an imperfect synchronisation of cell divisions.

In these preparations the level of contamination by the other cell compartments is not over 4 %. The absence of plasmalemma seems to be well established. When these vacuoles are taken out of the Ficoll used to constitute the density gradient bursting does not occur ; the vacuoles will maintain their integrity during at least 24 h at 4°C.

The main parameters influencing the quality of the preparations are : the duration of hydrolysis of the cell walls, the buffer composition, molarity and pH, the presence of 1 mM EDTA, the density and volume of the Ficoll layers, the quantity of deposited protoplasts and the duration of the ultracentrifugation.

METHOD 2 : VACUOLE ISOLATION USING DIGITONIN FOR PLASMALEMMA LYSIS

II. EQUIPMENT, BIOLOGICAL MATERIAL, CHEMICALS AND SOLUTIONS

A. Equipment, biological material

They are the same as in Method 1.

B. Chemicals

Mannitol ; 2(N-morpholino)ethanesulfonic acid (MES) ; tris (hydroxymethyl) aminomethan (Tris) ; ethylenediamine-te-traacetic acid (EDTA) ; Fluorescein diacetate (FDA) ; Conca-navalin A fluorescein labeled (all from Sigma), Ficoll 400 (Pharmacia), Macerozyme R 10 (Yakult Honsha Co., LTD, Tokyo), Cellulase Y-C, Pectolyase Y-23 (Sheishin Pharmaceutical Co., LTD, Tokyo) ; Digitonin (Sigma), Triton-X 100 (Sigma), Triethanolamine hydrochloride, Sodium phosphate, Magnesium chloride, Glucose-6-phosphate, NADP$^+$, 30 % Hydrogene peroxide, Cytochrome c, Antimycin, Sodium ascorbate, N,N,N',N'-Tetramethyl-p-phenylene-diamine (TMPD), Na-Citrate, p-nitrophenyl-α-D-mannoside, Na-Borate.

C. Solutions

1. Solutions for protoplast isolation

See Method 1.

2. Solutions for vacuole isolation

a) 0.7 M Mannitol, 5 mM Tris, 12 % (w/v) Ficoll, pH 7,9

b) 0.7 M Mannitol, 5 mM Tris, 6 % (w/v) Ficoll, pH 7,9

c) 0.7 M Mannitol, 5 mM Tris, pH 7.9

d) Digitonin solution : 1.5 % (w/v) digitonin is solubilized in a hot solution consisting of 0.7 M Mannitol, 5 mM Tris, 12 % (w/v) Ficoll, pH 7.9.

3. Solutions for the assay of marker enzyme activities

* glucose-6-phosphate dehydrogenase
          - Triethanolamine buffer, 0.1 M, pH 7.6
          - Magnesium chloride solution, 0.1 M
          - Glucose-6-P solution, 35 mM
          - NADP solution, approx. 11 mM (10 mg NADP$^+$, Na-salt/ml).

* Catalase

>   - Phosphate buffer, 50 mM, pH 7.4
>   - Buffer/H_2O_2-solution : add to 50 ml phosphate
> buffer 0.06 ml 30 % H_2O_2 and mix well, adjust the optical
> density of this solution against buffer (1) at 240 nm to
> 0.500 $\pm$ 0.010 (d = 1 cm).

* Cytochrome oxidase

>   - Phosphate buffer 50 mM, pH 7.4
>   - Oxidized cytochrome c, 50 mg.ml^{-1}
>   - Triton X-100 2.5 % (w/v)
>   - antimycin, alcoholic solution, 2.5 µg.$µl^{-1}$
>   - Na-ascorbate, 0.5 M
>   - TMPD (N,N,N',N'-tetramethyl-p-phenylene diami-
>   ne), 10^{-2} M

* α-mannosidase

>   - Citrate buffer, 0.1 M, pH 4.5
>   - p-nitrophenyl-α-D-mannoside 10^{-2}M
>   - Borate buffer, 0.2 M, pH 9.8.

III. EXPERIMENTAL PROCEDURE

8.45 am 1. Isolation of protoplasts

See Method 1.

10.00 am   2. Isolation of vacuoles

Vacuoles from mature cells are large organelles. Their
separation from other components of the cell is usually
performed by sedimentation in Ficoll gradients containing

isotonic mannitol (0.5 to 0.7 M). The purification of the
vacuoles obtained after treatment of protoplasts with digitonin by sedimentation in Ficoll layers was unsucessful. A light
aggregate was progressively formed after lysis. Vacuoles were
trapped within the network formed by this material and we were
unable to obtain a clean fraction. The chemical composition of
this undesirable aggregate is unknown. It probably corresponds
to that described by Wagner (1983), which was formed during
isolation of vacuoles from osmotically-lysed protoplasts.
Thus, vacuoles were alternatively separated by a flotation
procedure. For this purpose, the treatment of protoplasts with
digitonin was done in a viscous medium of high density (10.8 %
Ficoll, 0.7 M mannitol, 5 mM Tris, pH 7.9) at 0°C. Because
their density is lower than both the density of the medium and
that of other organelles, vacuoles released from protoplasts
after plasmalemma disruption float on the surface of the
reaction medium whereas other organelles as well as soluble
cytoplasmic enzymes (on account of the high viscosity of the
medium) remain in the bottom layer. Addition of a layer of
intermediate density allows a better separation of the
vacuolar fraction.

     The following experimental conditions were routinely
used : 1 ml protoplasts (about 20 millions cells) were
suspended in 5.5 ml 0.7 M mannitol, 5 mM Tris, 12 % (w/v)
Ficoll, pH 7.9, and supplemented with 3.5 ml of a digitonin
solution [1.5 % (w/v) digitonin solubilized in 0.7 M mannitol,
5 mM Tris, 12 % (w/v) ficoll, pH 7.9]. The mixture was allowed
to stand on ice for 5 minutes with occasional shaking. Then, a
layer (1 volume per volume of reaction medium) of 6 % (w/v)
Ficoll, 0.7 M mannitol, 5 mM Tris, pH 7.9 were pipetted onto
the lysate. Vacuoles were recovered at the interface 0 % - 6 %
Ficoll after 20 minutes centrifugation at 1.000 x g.

2.30 pm 3. <u>Control of the purity of the vacuole preparations</u>

Marker enzyme activities must be measured in both the protoplast and vacuole fractions. Prior to enzymatic determinations, protoplasts and vacuoles must be treated with Triton-X 100 (final conc. : 0.2 %) in 50 mM phosphate buffer, pH 7.4 at 0°C, in order to solubilize membranes.

3.1. Glucose-6-phosphate dehydrogenase

Measurement of the glucose-6-phosphate dehydrogenase activity is used to check the absence of cytoplasmic contamination.

Pipette into cuvette		Concentration in the assay mixture
Triethanolamine buffer	2.59 ml	86.3 mM
$MgCl_2$ solution	0.20 ml	6.7 mM
Glucose-6-P solution	0.10 ml	1.2 mM
NADP-solution	0.10 ml	0.37 mM

Mix, start reaction by addition of the sample

The absorbance at 340 nm is recorded
(ε_{340} = 6.22 cm² µmol^{-1})

3.2. Catalase

The activity of catalase is assayed in order to estimate the peroxisomal contamination.

Pipette into cuvette :		Concentration in assay
Buffer/H_2O_2 (II)	3.00 ml	50 mM phosphate 10.5 mM H_2O_2

Start reaction by addition of the sample

The absorbance at 240 nm is recorded ($\varepsilon_{240} = 0.040$ cm^2.µmol^{-1})

3.3. Cytochrome oxidase

The activity of cytochrome oxidase (inner mitochondrial membrane marker) is estimated by measuring oxygen consumption with an oxygen electrode.

Pipette into the thermostated cell		Concentration in the assay mixture
Phosphate buffer	0.5 ml	50 mM
Antimycin	2 µl	5 µg
Cytochrome c	20 µl	50 µM
Na-ascorbate	10 µl	3.3 mM
TMPD	5 µl	33 µM

Start the reaction by addition of the solubilized sample

4. Assessment of the efficiency of the method

The measurement of total activity of α-mannosidase, a specific marker of the vacuole of **A. pseudoplatanus** cells, can be used to estimate the yield of the isolation procedure.

Pipette into cuvette :		Concentration in the assay mixture
Citrate buffer	0.50 ml	16.5 mM
p-nitrophenyl-α-D-mannoside	0.50 ml	1.6 mM
Start reaction by addition of the sample		
Incubate 15 min.		
Stop reaction by addition of :		
Borate buffer	2.00 ml	13.2 mM
Mix, read optical density against blank at 405 nm		

IV. COMMENTS

The digitonin-induced lysis of the plasmalemma appears to be a convenient way to disorganize the plasmalemma without affecting the tonoplast. Vacuoles can be separated from other cell components by a flotation procedure. The yield of the isolation step and the purity of the fraction are quite satisfactory. The method is rapid, requires only simple equipment and allows the preparation of large amounts of vacuoles in a short time.

METHOD 3 : ISOLATION OF VACUOLES AFTER LYSIS
OF PROTOPLASTS BY OSMOTIC SHOCK

II. EQUIPMENT, BIOLOGICAL MATERIAL, CHEMICALS AND SOLUTIONS

A. Equipment

1. Access to

- a refrigerated centrifuge (up to 2000 g) with a swinging bucket rotor of 240 ml capacity (16 x 15 ml) e.g. Jouan E 96 N)
- a shaking bath with temperature control (10 to 80 rotations per min ; 28°C)
- a microosmometer
- a vacuum filtration system with fritted glass filters (40 to 90 µm)
- a scale
- a pH meter
- a standard microscope
- a magnetic stirrer with teflon-coated magnetic tods

2. On the bench

- glassware
- haemocytometers : Nageotte (50 µl content) for protoplasts counts and Malassez (1 µl content) for vacuole counts
- pyrex round bottom centrifuge tubes (15 ml capacity)
- plastic round bottom centrifuge tubes (30 ml capacity)
- plastic disposable pasteur pipettes
- nylon cloth (104 µm and 53 µm pore diameter)

- graduated glass pipettes (1 ml, 5 ml, 10 ml)
- automatic pipettes (5 µl to 5000 µl)
- a bucket with ice

B. Biological material : Acer pseudoplatanus isolated cells

C. Chemicals

Tris (hydroxymethyl)-aminomethan (Tris) ; 2(N-morpholino) ethanesulphonic acid (MES, Sigma Chemicals Co, Saint-Louis, USA) ; Na_2-EDTA (Prolabo, Paris, France) ; D-mannitol (Prolabo, Paris, France) ; $CaCl_2$, 2 H_2O (Prolabo, Paris, France) ; Ficoll 400 (Pharmacia, Uppsala, Sweden) ; enzymes : Pectolyase Y23 (Sheishin Pharmaceutical Co, Tokyo, Japan) ; Cellulase Onozuka RS (Yakult Honsha Co, Tokyo, Japan).

D. Solutions

25 mM Tris and 25 mM MES solutions in distilled H_2O are needed to make all solutions.

Solution 1 : for washing the cells and preparing the protoplasts : 25 mM Tris-MES, pH 5.5 containing 1 mM Ca^{2+}, final osmolality adjusted to 600 mosm with mannitol,

Solution 2 : for washing the protoplasts : 25 mM Tris-MES, pH 7.2, final osmolality adjusted to 600 mosm with mannitol,

Solution 3 : shock medium : 25 mM Tris-MES, pH 7.4 containing 2 mM Na_2-EDTA,

Solution 4 : stabilization medium : 25 mM Tris-MES containing 6 % Ficoll, pH 7.2 ; final osmolality adjusted to 550 mosm with mannitol,

Solution 5 : intermediate layer : 25 mM Tris-Mes, containing 3 % Ficoll, pH 7.2 ; final osmolality adjusted to 550 mosm with mannitol,

Solution 6 : top buffer for collecting the vacuoles :
25 mM Tris-MES, pH 7.2, final osmolality adjusted to 550 mosm
with mannitol.

All the solutions are stored sterile after autoclaving
at 120°C for 20 min.

III. EXPERIMENTAL PROCEDURE

1. Preparation of protoplasts

9.00 am

Cells (8 g fresh weight) at exponential phase of
growth were incubated in the enzyme solution (20 ml solution 1
containing 0.1 % pectolyase Y23 and 1 % cellulase RS) at 28°C,
in a water bath shaker (40 rotations per min).

10.30 am

Once most of the cell-walls have been digested and a
majority of cells have released their protoplast (after
1 h 30) ; the suspension is successively filtrated through
nylon cloth of 104 µm and 53 µm pore diameter and the
protoplasts collected in a plastic disposable tube (30 ml)
kept in ice. They are centrifuged for 5 min at 160 g at 4°C.
The supernatant is discarded and the pellet resuspended in 20
ml of ice-cold solution 2. The suspension is washed twice and
the final pellet is resuspended in 10 ml of the same buffer.
The number of protoplasts per ml is determined using a
Nageotte haematocytometer ; the protoplast concentration is
adjusted to 4 to 6 x 10^6 protoplasts per ml.

2. Osmotic shock and preparation of vacuoles

10.45 am

An erlenmeyer flask of 125 ml is used. One ml of
protoplast suspension is added to 900 µl of solution 3 and
gently agitated at 26°C (by rotation). When most of the
protoplasts have burst releasing their vacuoles (after 12
min), 12 ml of ice-cold solution 4 are gently added in the
suspension to stabilize the vacuoles ; the flask is constantly

stirred by hand to homogenize the ficoll. The cytoplasmic debris (plasma membrane, mitochondria, starch granules...) can then be easily pipetted out of the suspension as they accumulate in the middle of the suspension forming a pellet. The suspension is poured in a pyrex centrifuge tube (15 ml capacity) and overlayed with 1 ml of solution 5 containing Ficoll 3 % and one ml of solution 6. The gradient is centrifuged for 10 min (4°C) at 1500 g. The vacuoles are recovered at two levels in the gradient (Fig. 1 a) : at the top (V1) and at the interphase between solution 6 and Ficoll 3 % (V2). They are collected (600 µl per gradient) with a plastic disposable pipette and counted in a Malassez haematocytometer.

The time needed to go through the whole procedure for preparing 1 to 10 ml vacuole suspension is about **2 h 30.**

3. Control of the purity of the vacuole preparations
2.30 pm

(See Methods 1 and 2)

IV. COMMENTS

In this paragraph we give some advice and comments concerning the preparation of the vacuoles and a discussion about the possible origin of vesicles which exist in vacuole suspensions.

The yield of protoplasts and vacuoles is directly influenced by the physiological state of the cells which have to be in the middle of the exponential phase of growth ; in these conditions, the yield of protoplasts compared to cells is 60-70 %. The vacuoles are recovered at two different levels on the gradient (fig. 1 a) in a ratio of about 20-25 % (for each level) of the amount of protoplasts deposited in the shock medium. In V1, the contamination with protoplasts is very low (less than 1 %) ; as observed under light microscope the suspension is clean, not contaminated with membranes and

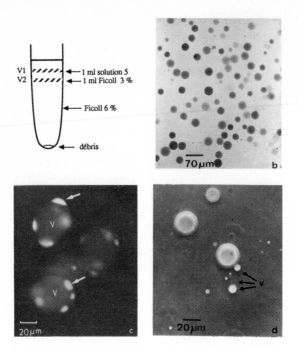

Fig. 1

a) Schematic representation of a gradient used for the preparation of vacuoles isolated from **Acer pseudoplatanus** cells.

After centrifugation, the vacuoles are recovered at two interphases, V1 and V2.

b) Suspension of vacuoles isolated from **Acer pseudoplatanus** cells and stained with neutral red.

The population is heterogenous in size and in the intensity of the staining.

c) Micrograph of **Acer pseudoplatanus** cells which have absorbed fluorescein.

The cells were incubated at pH 6.0 in the presence of fluorescein (3 μm) for 15 min. They show small fluorescent vesicles (arrows) surrounding the large central vacuole V.

d) Example of the release of vesicles v, during the stabilization of a suspension of vacuoles from **Catharanthus roseus** cells when the protoplasts are submitted to an osmotic shock (from P. Manigault).

The double membrane layer which outlines the vacuoles is an artefact, due to the fact that only the small vesicles have been brought into focus.

starch granules. The V2 suspension is only slightly contaminated ; the major problem with this layer is that the vacuoles
are in direct contact with the stabilization buffer making it
difficult to pipet out the vacuoles without contamination. In
any case, it is necessary to check the purity of a vacuole
suspension using marker enzymes to detect the presence of
endoplasmic reticulum, mitochondria, nuclear membrane, cytosol
and microbodies. The level of contamination should not exceed
1 - 2 %.

The time needed for osmotic shock to be completed
depends, at least in part, on the stability of the protoplast
suspension which varies with the physiological state of the
cells. As a consequence, the protoplast lysis has to be
carefully watched and the process stopped when not more than 1
to 3 % of protoplasts remain intact in the preparation.

The presence of EDTA in the stabilization buffer at
neutral pH helps the debris not to stick on the vacuoles which
are just released from the protoplasts but to aggregate,
facilitating their isolation from the suspension before
centrifugation.

The pH of the upper layer is also critical as the
vacuoles show the maximum of stability at pH around 7.0 - 7.5.

Ficoll has been preferred over nycodenz : the rate of
its diffusion into buffer solutions is lower, making it easy
to build gradients with interphases which remain clear even
after centrifugation.

What is the possible origin of the vesicles ?

Vesicles appearing in a vacuole suspension can origi-
nate at least from two sources. It is well known that a
vacuole suspension is heterogenous in size (fig. 1 b) and
contains a variable percentage of small vacuoles (vesicles
with a diameter of 5 - 10 μm). The mean diameter of the
vacuoles is about 25-30 μm. This heterogeneity in size is only
partly due to the variability of the size of the protoplasts
from which the vacuoles are issued. Some cell species such as

Acer contain a large central vacuole and a variable number of vesicles (1 to 12) (Alibert et al., 1982) (fig. 1 c). Some of these vesicles remain enclosed in the membranes during protoplast lysis and sediment at the bottom of the tube during centrifugation while other float and are recovered in the upper layers of the gradient together with the vacuoles. This phenomenon is however, not the only one leading to the appearance of vesicles. They can also be issued from the central vacuoles during their isolation. The step of Ficoll addition (or nycodenz in the case of **Catharanthus** vacuoles) is critical. It produces a sudden increase of the osmotic pressure of the medium which creates an "hyperosmotic" shock to the vacuoles ; these in turn readjust their internal pressure by releasing one or several small vesicles (fig. 1 d). A variable quantity of these vesicles are recovered at levels V1 or V2 ; the majority of them remain in the stabilization buffer. The phenomenon of vesicle formation has to be taken into account when studying ion exchanges across the tonoplast because of the low surface to volume ratio of vacuoles compared to that of vesicles.

The mean number of vacuoles per protoplast when calculated from the data obtained with α-mannosidase as a vacuolar marker is about 2 (Alibert et al., 1982) ; for this reason it is important to use such a marker to calculate the concentration of a vacuolar solute in the protoplasts.

In conclusion, most of the techniques which have been developed these recent years permit the isolation of vacuoles which are in "good physiological conditions", i.e. able to transport ions and metabolites. However, when studying transport mechanisms across the tonoplast, one has to keep in mind that an isolated vacuole does not behave like a vacuole **in situ** : the tonoplast is not energized and the organelle is not integrated in the functioning of the whole cell in which constant exchanges occur between the cytoplasm and the vacuole sap.

V. REFERENCES

Admon A., Jacoby B., Goldschmidt E.E. (1980). Assessment of cytoplasmic contaminations in isolated vacuole preparations. **Plant Physiol. 65** : 85-87.

Alibert G., Carrasco A., Boudet A.M. (1982). Changes in biochemical composition of vacuoles isolated from **Acer pseudoplatanus** L. during cell culture. **Biochim. Biophys. Acta 721** : 22-29.

Boller T., Dürr M., Wiemken A. (1976). Asymetric distribution of concanavalin A binding sites on yeast plasmalemma and vacuolar membrane. **Arch. Microbiol. 109** : 115-118.

Boller T., Kende H. (1979) Hydrolytic enzymes in the central vacuole of plant cells. **Plant Physiol. 63** : 1123-1132.

Boller T., Wiemken A. (1986). Dynamics of vacuolar compartmentation. **Ann. Rev. Plant Physiol. 37** : 137-164.

Boudet A.M., Canut H., Alibert G. (1981). Isolation and characterization of vacuoles from **Melilotus alba** mesophyll. **Plant Physiol. 68** : 1354-1358.

Buser C., Matile Ph. (1977). Malic acid in vacuoles isolated from **Bryophyllum** leaf cells. **Z. Pflanzenphysiol. 82** : 462-466.

Chanson A., Mc Naughton E., Taiz L. (1984). Evidence for a KCl-stimulated, Mg^{2+}-ATPase on the Golgi of corn coleoptiles. **Plant Physiol. 76** : 498-507.

Doll S., Rodier F., Willenbrink J. (1979). Accumulation of sucrose in vacuoles isolated from red beet tissue. **Planta 144** : 407-411.

Grob K., Matile Ph. (1979). Vacuolar location of glucosinates in horseradish root cells. **Plant Sci. Lett. 14** : 327-335.

Hartmann-Bouillon M.A., Benveniste P. (1978). Sterol biosynthetic capability of purified membrane fractions from maize coleoptiles. **Phytochemistry 17** : 1037-1042.

Leigh R.A. (1983). Methods, progress and potential for the use
    of isolated vacuoles in studies of solute transport in
    higher plant cells. **Physiol. Plant. 57** : 390-396.
Leigh R.A., Branton D. (1976). Isolation of vacuoles from root
    storage tissue of **Beta vulgaris** L. **Plant Physiol. 58** :
    656-662.
Le Quôc K., Le Quôc D., Pugin A. (1987). An efficient method
    for plant vacuole isolation using digitonin for plasmalemma
    lysis. J. **Plant Physiol. 126** : 329-335.
Lorz H., Harms C.T., Potrykus I. (1976). Isolation of "vacuo-
    plasts" from protoplasts of higher plants. **Biochem.
    Physiol. Pflanz. 169** : 617-620.
Matile Ph. (1968). Lysosomes of root tip cells in corn seed-
    lings. **Planta 79** : 181-196.
Matile Ph. (1978). Biochemistry and function of vacuoles. **Ann.
    Rev. Plant Physiol. 29** : 193-213.
Martinoia E., Heck U., Wiemken A. (1981). Vacuoles as storage
    compartments for nitrate in barley leaves. **Nature 289** :
    292-293.
Marty F., Branton D. (1980). Analytical characterization of
    beetroot vacuole membrane. J. **Cell. Biol. 87** : 72-83.
Pugin A., Montrichard F., Le-Quôc K., Le-Quôc D. (1986).
    Incidence of the method for the preparation of vacuoles on
    the vacuolar ATPase activity of isolated **Acer pseudoplata-
    nus** cells. **Plant Sci. Lett. 47** : 165-172.
Pujarniscle S. (1968). Caractère lysosomal des lutoïdes de
    latex d'**Hevea brasiliensis. Physiol. Veg. 6** : 27-46.
Quail P.H. (1979). Plant cell fractionation. **Ann. Rev. Plant
    Physiol. 30** : 425-484.
Schmidt R., Poole R.J. (1980). Isolation of protoplasts and
    vacuoles from storage tissue of red beet. **Plant Physiol.
    66** : 25-28.

Schumaker K.S., Sze H. (1985). A Ca^{2+}/H^+ antiport system driven by the proton electrochemical gradient of a tonoplast H^+-ATPase from oat roots. **Plant Physiol.** **79** : 1111-1117.

Wagner G.J. (1983). Higher plant vacuoles and tonoplasts. In : Hall J.L., Moore A.L. (eds) Isolation of membranes and organelles from plant cells, Academic Press London pp 83-117.

Wagner G.J., Siegelman H.W. (1975). Large-scale isolation of intact vacuoles and isolation of chloroplasts from protoplasts of mature plant tissues. **Science 190** : 1298-1299.

Experiment nº 7

ASSAY OF IN VITRO MICROTUBULE ASSEMBLY.
EFFECT OF MICROTUBULE ASSOCIATED PROTEINS

F. PIROLLET & D. JOB

I. INTRODUCTION AND AIMS

Microtubules are fibrous elements, made of a dimeric protein (tubulin), found ubiquitously in eucaryotic cells where they perform a wide variety of physiological functions. They are centrally involved in mitosis and other motility related phenomena such as the transport of subcellular organelles or changes in cell shape (Snyder & Mc Intosh, 1976 ; Dustin, 1978 ; Margolis & Wilson, 1981).

To perform these tasks, they must be able to assemble at the proper time and place and with the proper orientation in response to physiological signals. It is thus reasonnable to assume that microtubules associated machineries with a regulatory function must exist in cells. Their characteriza- tion relies to a large extent on the study of **in vitro** tubulin assembly and its modification by various tissue extracts.

The classical method to isolate microtubules from crude extracts is to assemble these polymers and to perform subsequently three cycles of depolymerisation in the cold, followed by a new assembly step.

The final preparation contains 75 % tubulin and 25 % microtubule associated proteins (MAPs). This is the simplest system in which tubulin assembly can be shown to be regulated by other proteins whose function can in turn be modified in response to physiological signals.

The experiments that are proposed below describe some basic methods that have been used to investigate tubulin assembly and its regulation by microtubule associated pro- teins.

II. EQUIPMENT, CHEMICALS, SOLUTIONS

A. Equipment

1. Access to

- a cold room (4°C)
- a liquid-scintillation counter
- a spectrophotometer equipped with a constant temperature chamber as UVIKON 810
- electron microscopy facilities
- a power supply for electrophoresis

2. On the bench

- glassware, automatic pipettes
- disposable plastic tubes (5 ml)
- filter GF/C purchased from Whatman (ref 1822025)
- filters holder with a funnel connected to a vacuum flask supplied from Millipore (XX 1 025 00)
- chromatography columns (Pharmacia K9x15) with plastic tubing and connections
- a polyacrylamide gel slab cell (biorad : miniprotean II)
- distributors for 1,4 and 9 ml
- clocks

B. Chemicals

- Microtubule protein from beef brain isolated by three cycles of assembly and disassembly, in MME buffer, according to published procedures (Asnes & Wilson, 1979 ; Job & Margolis, 1984) and stored at - 80°C for later use.
- Phosphocellulose P11 (Whatman) precycled prior to use as indicated by the manufacturer.

- Acetylphosphate, GTP, acetate kinase (101834), ATP from Boehringer Mannheim.
- 2-(N-morpholino)-ethane sulfonic acid (Mes), ethylene glycol bis (β-aminoethyl ether)-N, N,N',N',-tetraacetic acid (EGTA), from Sigma.
- Sucrose, glutaraldehyde 25 % from Merck.
- Uranylacetate, electron microscopy grids (G 200) from Polaron.
- ^{3}H-GTP (TRK 314) from Amersham.

C. Solutions

- MME buffer : Mes 100 mM, $MgCl_2$ 1 mM, EGTA 1 mM, pH : 6.75
- 50 % sucrose - MME buffer : addition of sucrose 50 % (w/v) to previously prepared MME.
- Washing buffer : 10 % DMSO - 25 % glycerol - H_2O
- DGMME buffer : MME buffer made 5 % DMSO, 10 % glycerol, 15 mM $MgCl_2$
- Assembly mixture for filter assay : 20 mM acetyl phosphate, 1 µg/ml acetate kinase, 0,1 mM GTP-^{3}H (40 µCi/ml) in DGMME buffer
- Solutions for polyacrylamide gel electrophoresis under denaturing conditions (SDS-PAGE) according to Laemmli & Favre (1976).

The solutions will be prepared in advance, kept at 4°C except the assembly mixture.

III. EXPERIMENTAL PROCEDURES

Separation of pure tubulin from MAPs using phosphocellulose chromatography

Microtubule protein(7.5 mg) is loaded onto a 3 ml phosphocellulose column in MME buffer, using a Pharmacia K9x15 column. The column is eluted with the same buffer and fractions of 600 µl are collected. Pure tubulin is eluted mainly in the fourth fraction. MAPs are absorbed onto the ion exchanger.

Turbidimetric assay of microtubule assembly

The samples (300 µl) to be studied are merely put into a microcuvette at 30°C in the spectrophotometer. After adjustement of the baseline, 1 mM GTP is added and the O.D. variations at 350 nm are monitored. At steady-state, the temperature is shifted to 6°C and microtubule disassembly is monitored similarly.

This turbidimetric method will be applied to 3 different samples : total microtubule protein, pure tubulin in MME buffer and pure tubuline in DGMME buffer, all samples at 1,5 mg/ml.

Assay of microtubule associated proteins

The MAP-assay relies on a radioactivity labeling method of assembled microtubules (Wilson et al., 1982). The original description of the MAP-assay is published by Job et al. (1985). During tubulin assembly GTP is incorporated into the microtubule wall. Using ^{3}H-GTP one can quantitatively assay the amount of assembled tubulin by measuring the corresponding radioactivity. The labeled microtubules can be separated from free subunits and from unincorporated radioac- tivity by filtration on GF/C filters.

Phosphocellulose purified tubulin (PC-tubulin), (500 µl at 4 mg/ml) is diluted with an equal volume of the above defined assembly mixture. Four 40 µl aliquots are immediatly diluted in 1 ml of 50 % sucrose - MME containing 0.75 % glutaraldehyde (blanks).

The rest of the mixture is incubated for 50' at 30°C. It is then aliquoted in 18 tubes, containing 1 ml of 50 % sucrose - MME buffer incubated at 30°C and containing :

- 0.75 % glutaraldehyde (4 tubes) : total assembly level controls. These samples are then left at room temperature before further processing.

- Variable amounts of MAPs added as unassembled microtubule protein. Seven MAPs concentrations will be tested in duplicate : 0 - 0.05 - 0.1 - 0.2 - 0.3 - 0.5 - 1. These concentrations are expressed as the ratio of the amount of total microtubule protein as total pure tubulin in the assay. In practice, total microtubule protein is diluted to 2 mg/ml. Aliquots of 0, 2, 4, 8, 12, 20 and 40 µl are added to the various tubes.

The samples are incubated for 30 min. at 30°C. They are then diluted to 30 % sucrose by addition of prewarmed MME buffer (666 µl). Ten minutes later, 30 µl of 25 % glutaraldehyde are added. The samples are then left at room temperature.
The filter assay itself is performed on a filter holder. GF/C filters are preequilibrated in MME buffer containing 3 mM ATP and then placed on the apparatus. The filter is washed once with 4 ml of washing buffer. The sample to be analysed is then poured onto the filter. Two more 4 ml washes are performed.
The filters are transfered in scintillation vials, and incubated 10 min. with 1 ml of EtOH. 9 ml of scintillation liquid are added and the radioactivity is counted.
PC-tubulin is unstable to dilution in 30 % sucrose - MME. The corresponding filters will show a level of radioactivity equivalent to blank levels. MAPs effect is manifested by the induction of a resistance of microtubules to similar dilution conditions with a corresponding increase in the radioactivity levels trapped on filters.

Other methods

Slab gel electrophoresis is run in 8 % polyacrylamide/0.1 % sodium lauryl sulfate, according to Laemmli & Favre (1983). Gels are stained with Coomassie blue R.

For electron microscopy study of the microtubules, samples are diluted 40 fold in 50 % sucrose - MME buffer containing 0.75 % glutaraldehyde. This material is allowed to adhere on a formwar 0.5 % - coated E.M. grid. The grid is then negatively stained with 1 % uranyl acetate.

General outlook of the day

Morning 8.45 am

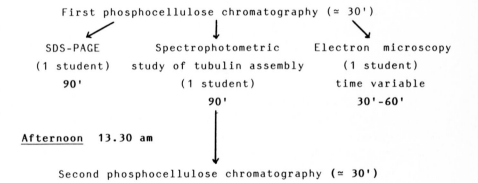

First phosphocellulose chromatography ($\simeq$ 30')

SDS-PAGE	Spectrophotometric	Electron microscopy
(1 student)	study of tubulin assembly	(1 student)
90'	(1 student)	time variable
	90'	30'-60'

Afternoon 13.30 am

Second phosphocellulose chromatography ($\simeq$ **30'**)

MAPs assay
(all attendents)
(3 hours)

IV. REFERENCES

Asnes C.F. and Wilson L. (1979). Isolation of bovine brain amicrotubule protein without glycerol : polymerisation kinetics change during purification cycles. **Anal. Biochem.,** **98** : 64-73.

Dustin P. (1978). Microtubules (Springer Verlag, New-York).

Job D. and Margolis R.L. (1984). Isolation from bovine brain of a superstable microtubule subpopulation with microtubule seeding activity. **Biochemistry, 23** : 3025-3031.

Job D., Pabion M. and Margolis R.L. (1985). Generation of microtubule stability subclasses by microtubule-associated proteins : implications for the microtubule "dynamic instability" model. **J. Cell. Biol., 101** : 1680-1689.

Laemmli I.K. and Favre M. (1973). Maturation of the head of bacteriophage T_4 : I. DNA packaging events. **J. Mol. Biol., 80** : 575-580.

Margolis R.L. and Wilson L. (1981). Microtubule treadmills - possible molecular machinery. **Nature** (London) **292** : 705-711.

Snyder J.A. and McIntosh J.R. (1976). Biochemistry and physiology of microtubules. **Ann. Rev. Biochem., 45** : 699-720.

Wilson L., Snyder K.B., Thompson W.C. and Margolis R.L. (1982). A rapid filtration assay for analysis of microtubule assembly, disassembly, and steady-state tubulin flux. **Methods Cell. Biol., 24** : 159-169.

Experiment n° 8

HISTAMINE RELEASE FROM HUMAN BASOPHILES

A. MOREL and M. DELAAGE

I. INTRODUCTION AND AIMS

Histamine, β imidazolethylamine (MW = 111) is an intercellular mediator exerting potent effects on various target tissues.

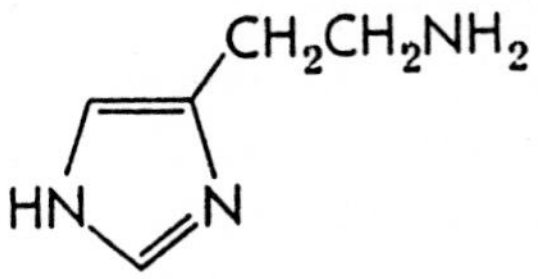

Histamine

It is produced by decarboxylation of histidine.

Ubiquitously distributed in mammalian tissues, histamine is one of the major mediators of immediate hypersensitivity reactions (Siragagnian, 1983 & Wasserman, 1983) : stored in large amounts in metachromatic granules of mastocytes and basophils, it is actively released from these cells when challenged with allergens (to which the allergic patient is sensitized) or with "non specific" histamine releasers such as anti-IgE or pharmacological agents. Histamine is also an important mediator in the central nervous system and the gastric mucosa.

The clinical interest of histamine determination is in the establishment of liable agents for immediate hypersensitivity reactions. These allergens may be pointed out either **in** vitro using allergen histamino-liberation from basophils (Lichtenstein & Osler, 1964) or **in vivo** : it is then possible to measure the variation of histamine levels in fluids such as plasma (Bhat et al., 1976 ; Bruce et al., 1976), urine (Bruce et al., 1976) or bronchial lavage fluids (Butchers et al., 1980) after challenge with allergens. In pregnant women, the histamine catabolism is increased.

For research works, the quantification of histamine is of major interest. If the pharmacological effects of histamine (Douglas, 1975) on smooth muscles (tremendous during anaphylactic shocks) or on gastric secretion seem to be fairly well known, histamine could have several other roles to play through H_1 or H_2 receptors in the central nervous system (Schwartz, 1979) as well as at the peripheral level (immunomodulation : Beer et al., 1984, cell growth : Kahlson & Rosengreen, 1971).

PRINCIPLE OF HISTAMINE DETERMINATION

Immunotech has developped a kit containing tubes coated with a highly specific monoclonal antibody and reagents ready to use for the histamine acylation essential for a good antibody recognition. This conversion is fast and complete.

The histamine determination is based on the competition of acylated histamine and I^{125}-acylated histamine for their binding to the mouse monoclonal antibody fixed on a tube.

II. EQUIPMENT AND CHEMICALS

A. Equipment

Beside usual laboratory equipment, the following equipment is required :

1. For the radioimmunoassay

- precision micropipettes (50, 100 and 500 µl)
- gamma counter
- device for aspiration
- shaker such as vortex
- a cold room (+ 4°C)

2. For histamine release

- 96 U shaped wells microtiter plate
- fixed volume multipette and combitips (5 ml)
- 37°C oven
- refrigerated centrifuge for microtiter plates

B. Chemicals (KIT supplied by Immunotech)

a.	^{125}I-acylated histamine	1 x 55 ml
b.	antibody coated tubes	100 tubes
c.	acylating reagents	50 tubes
d.	histamine standards	7 x 1 ml
e.	acylation buffer	1 x 5 ml
f.	histamino-liberation buffer	1 x 30 ml

a. ^{125}I-radiolabelled histamine

The vial contains 4 µCi or 148 KBq (at the date of manu-facture) of ^{125}I-acylated histamine diluted in 50 mM Borate buffer pH 8.3) containing BSA with sodium azide and a red dye.

To be stored at 2-8°C away from bright light.

b. Anti-histamine antibody coated tubes

To be stored at 2-8°C.
Stable until the expiry date mentioned on the label.

c. Acylating reagent

Each tube contains 1 mg of lyophilized acylating reagent.
To be stored in a dry place.
Stable until the expiry date mentioned on the label.

d. Standards

The 7 vials of histamine standards contain : 0, 0.5, 1.5, 5, 15, 50, 140 nM solutions for standard curve setting.
To be stored at 2-8°C. For a long conservation time standards must be stored at - 20°C.
When reused they must be completely thawed and vortexed.
Stable until the expiry date mentioned on the label.

e. Acylation buffer

Borate buffer   pH 8.2

f. Histamino-liberation buffer (lyophilized)

Add to the vial 30 ml of distilled water.
Reconstituted buffer should be kept at - 20°C.

III. EXPERIMENTAL PROCEDURE

<u>Day 1</u>
9.00 am

A. Sample collection

Collect 1 to 2 ml of blood in an heparinized tube exclusively. If the histamine release cannot be performed immediately, store the blood sample at 4°C (for 24 h maximum).

10.00 am

B. Procedure for histamine release

a. Whole blood is diluted to 1/4 in "Histamino-liberation" buffer.

b. 50 µl of various dilutions of histamine releaser agent (allergen, anti-IgE, ionophore...) in histamino-liberation buffer are distributed in the wells of the microtiter plate. Then, add 100 µl (with multipette) of whole blood diluted to 1/4.
Note : do not forget spontaneous histamine release control. In this case, dilution buffer is used instead of histamine releaser agent dilution.

c. Incubate the lid covered plate for 30 mn at 37°C. Centrifuge in the cold (4°C) for 5 mn at 1 500 RPM. Be sure to keep then the temperature of the plate underneath 20°C.

d. Total histamine is quantified as follows :
- Add 50 µl of non diluted blood sample to 1 ml of distilled water.

- Freeze the tube.
- After thawing, the tube is centrifuged and the supernatant is collected.

REMARKS :
Supernatants can be stored frozen if the radioimmunoassay is delayed.

C. Radioimmunoassay
11.00 am
For the acylation of samples and standards proceed as follows :

a. Collect the powder down to the bottom of the polypropylene tubes by gentle tapping on the table.

b. Proceed then to acylation tube after tube :
- Add to the powder 100 µl of standard solution or sample,
- Then add 50 µl of acylating buffer ready for use,
- Recap the tube, vortex it immediately until complete solubilisation of the reagent (even traces remaining on tap).

The reaction is complete when all the lyophilisate is dissolved.

 c. Vortex and transfer the mixture in coated tubes
as indicated on the following table :

Tube n°	Tracer	Acylated standard (nM)	Acylated sample
1, 2	0.5 ml	0.05 ml (0)	-
3, 4	0.5 ml	0.05 ml (0.5)	-
5, 6	0.5 ml	0.05 ml (1)	-
7, 8	0.5 ml	0.05 ml (5)	-
9, 10	0.5 ml	0.05 ml (15)	-
11, 12	0.5 ml	0.05 ml (50)	-
13, 14	0.5 ml	0.05 ml (150)	-
15, 16	0.5 ml	-	0.05 ml (X)
etc.	0.5 ml	-	0.05 ml (Y)

1.00 pm
 d. In two tubes, add 0.5 ml of ^{125}I-histamine
tracer. Cap the tubes and keep them until counting. Note T,
total cpm.
 e. Incubate all tubes for 18 hours at 4°C.

 Day 2
8.00 am
 f. Aspirate the tubes maintening them at 4°C. Be
sure all fluid is drawed up.
 g. Count all emptied tubes. Note B, bound cpm.

D. Results

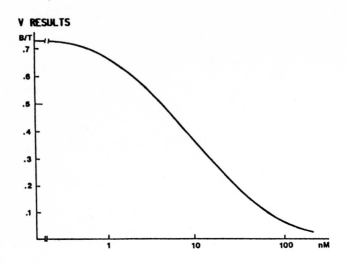

(standard curve given as an example ; do not use if for
calculation).

        Results are deduced by interpolation on standard
curve.

        a) Draw standard curves on linear graph paper
plotting :
                - on the horizontal axis, histamine concentration
of standards,
                - on the vertical axis, the average of the relative
radioactivity bound to tubes (B/T) or B/Bo.

        b) Results
                - Plasma histamine : direct reading on standard
curves.

- Histamine released : in order to relate the results to undiluted blood volume, multiply the result on standard curve by 6.
- Total histamine of whole blood (cell lysate in the hypotonic conditions and freezing) multiply the result by 21.

IV. COMMENTS

A. Specific assay characteristics

a. Sensitivity

The acylation allows a sensibility of the assay to 0.1 nM (in the sample).

b. Specificity

Cross-reactions with t-methylhistamine or histidine are very low ($< 10^{-4}$).

B. Limitation of the procedure

- For high histamine values, dilute the sample in order to read the result in the middle of the standard curve.

- Histamine is a common contaminant of biological material. The introduction of dusts or debris in the samples and reagents must be avoided. It is recommended to handle the kit with gloves.

V. REFERENCES

Beer D.J., Matloff S.M., Rocklin R.E. (1984). The influence of histamine on immune and inflammatory responses. **Adv. Immunol.**, **35** : 209-268.

Bhat K.N., Arroyave C.M., Marney S.R., Stevenson D.D., Tan E.M. (1976). Plasma histamine changes during provoked bronchospasm in asthmatic patients. **J. Allergy Clin. Immunol.**, **58** : 647-656.

Bruce C., Weatherstone R., Seaton A., Taylor W.H. (1976). Histamine levels in plasma, blood and urine in severe asthma and the effect of corticosteroid treatment. **Thorax**, **31** : 724-729.

Butchers H.R., Vardey C.J., Skidmore I.F., Wheeldom A., Boutal L.E. (1980). Histamine containing cells from bronchial lavage of macaque monkeys time course and inhibition of anaphylactic histamine release. **Archs. Allergy appl. Immun.**, **62** : 205-212.

Douglas W.W. (1975). Histamine and antihistamines ; 5-hydroxytryptamine and antagonists. **The pharmacological basis of therapeutic** (5ème ed.). Ed. Goodman L.S., Gilman A.G., Mac Millan publishing Co Inc, N.Y., 590-629.

Kahlson G., Rosengreen E. (1971). **Biogenesis and physiology of histamine.** Ed. H. Davson, A.D.M. Greenfield, R. Whittam, G.B. Brindley.

Lichtenstein L.M., Osler A.G. (1964). Studies on the mechanisms of hypersensitivity phenomena IX. Histamine release from human leucocytes by ragweed pollen antigen. **J. Exp. Med.** : 507-530.

Lorenz W., Neugebauer E., Schmal A. (1982). Le dosage de l'hista-mine plasmatique lors de réactions anaphylactoïdes chez le sujet anesthésié. Influence des méthodes de prélèvement et de la préparation du plasma sur l'histaminémie mesurée. **Ann. Fr. Anesth. Réanim.**, **1** : 271-276.

Morel A., Delaage M. - Brevet Immunotech n° 84.05783.

Schwartz J.C. (1979). Minireview : Histamine receptors in
brain. **Life Sci.**, **25** : 895-912.

Siraganian R.P. (1983). Histamine secretion from mast cells
and basophils. **TIPS, 4** : 432-437.

Wasserman S.I. (1983). Mediators of immediate-hypersensitivi-
ty. **J. Allergy Clin. Immun.**, **72** : 101-115.

Experiment n° 9

INTRACELLULAR Ca^{2+} MEASUREMENTS IN INTACT NEUTROPHILS

S. TREVES, D. MILANI & T. POZZAN

I. INTRODUCTION AND AIMS

During the past few years the importance of changes in the intracellular Ca^{2+} concentration $[Ca^{2+}]i$, in the activation of cellular functions has become more and more evident (Campbell, 1983). It is now generally accepted that stimulation of those membrane receptors linked to polyphosphonositide turnover leads to a rise in $[Ca^{2+}]i$ which is due both to the release of Ca^{2+} from vesicular intracellular stores and to increase influx via the plasma membrane (Berridge and Irvine, 1984). The initial sequence of events leading to cell activation through this signalling pathway can be summarized as follows : binding of a ligand to its receptor causes the activation, via a G protein (in some systems sensitive to pertussis toxin), of a specific membrane phosphodiesterase, also called phospholipase C (Di Virgilio et al., 1985). This enzyme hydrolyses phosphatidylinositol bisphosphate, PIP2, to inositol 1, 4, 5 trisphosphate, InsP3 (Berridge, 1984), (which release Ca^{2+} from vesicular intracellular stores) and diacylglycerol, DG, the physiological activator of protein kinase C (Nishizuka, 1984). The mechanism of Ca^{2+} influx across the plasma membrane is still unknown (Andersson et al., 1987).

Because of the importance of changes in the $[Ca^{2+}]i$ during cell activation, numerous methods to measure the concentration of this ion in living cells have been developed in the past few years. Compared to other techniques, fluorescent tetracarboxylates (quin2, fura-2 and indo-1) (Grinkiewicz et al., 1985) offer many advantages and have gained widespread use in many laboratories throughout the world. The basic principle underlying $[Ca^{2+}]i$ measurements with tetracarboxylates is straight forward. The acetoxymethyl ester (AM) form of the dye passes through the cell membrane and once inside the cell is hydrolyzed by cytoplasmic esterases, to the membrane impermeant free acid (FA) form. Thus simply by incubating the cells with μmolar concentrations of the AM tetracarboxylate ester derivative, one can trap large amounts of dye inside cells. Variations in the [Ca^{2+}]i can be monitored by following changes in the dye fluorescence. Different calibration procedures can be used in order to quantitate fluorescence values in terms of [Ca^{2+}] (see below). Tetracarboxylates vary in their affinity for Ca^{2+}, fluorescence and emission spectra, extinction coefficients and quantum yields (Grinkiewicz et al., 1985). Since we will use quin2 and fura-2, only the characteristics of these two indicators will be discussed.

Quin2 has a Kd for Ca^{2+} of 115 nM under conditions mimicking mammalian cell cytoplasm (100 mM KCl, 1 mM MgCl$_2$, pH 7.05, 37°C), peaks at ≃ 340, 500 nm (excitation and emission, respectively). Increases in [Ca^{2+}]i mainly result in increases in fluorescence intensity. Since the intracellular concentrations of quin2 which are necessary to swamp cell autofluorescence are of the order of several hundreds of μmoles/liter of cell water, this indicator greatly increases the soluble intracellular Ca^{2+} buffering capacity.

Fura-2 has been synthetized more recently by Grinkiewicz et al. (1985). The most interesting characteristic of this new indicator is its greater fluorescence (approximately 30 times on a per molar basis) compared to its precedessor quin2. Furthermore fura-2 responds to [Ca^{2+}]i changes not only with an increase in fluorescence intensity, but also with a shift in excitation spectrum. The affinity for Ca^{2+} of fura-2 is slightly lower than that of quin2 (Kd = 225 nM). Unfortunately there are several problems with fura-2 which have not been solved (leakage from intact cells, intracellular distribution etc.), so that its use has not become widespread yet.

In this section we will focus on the application of tetracarboxylates to measurements of $[Ca^{2+}]i$ in human neutrophils.

II. EQUIPMENT, CHEMICALS AND SOLUTIONS

A. Equipment

- Waterbath
- Swinging bucket centrifuge
- Spectrofluorimeter equipped with a thermostated cuvette holder and magnetic stirring
- Good quality glass cuvettes and small magnets
- Microfuge tubes
- Automatic pipettes (1-1000 µl)

B. Chemicals

Sodium citrate, NaCl, KCl, Na_3PO_4, HEPES (N-2 hydroxymethyl piperazine-N'-2-ethane sulfonic acid, $MgSO_4$, Glucose, Dextran T500 (Pharmacia, Uppsala, Sweden), Ficoll Hypaque (Pharmacia), Bovine Serum Albumin (BSA), quin2/AM (Calbiochem), fura-2/AM (Calbiochem), EGTA (ethyleneglycol-bis-(2-aminoethyl) tetra acetic) acid (Fluka), Tris (tri-(hydroxymethyl)aminomethane), $CaCl_2$, Dimethylsulfoxide (DMSO, Sigma), Triton-X 100 (Packard), formyl-Methionine-Leucine-Phenylalanine (fMLP, Sigma), Phorbol 12-myristate 13-acetate (PMA, Sigma), ionomycin (Calbiochem).

C. Solutions

For the preparation of human neutrophils :

a) 50 ml fresh blood

b) Sodium citrate (as anticoagulant). The stock citrate solution contains 31.3 g/l in 0.9 % NaCl and 20 mM HEPES, pH 7.4 and is added to the blood at 1:10 dilution (vol/vol).

c) Dextran T500 (Pharmacia, Uppsala, Sweden) is made 4 % in 0.9 % NaCl (use 1:5 vol/vol).

d) A small amount of 18 % NaCl

e) Ficoll Hypaque

f) A stock solution of the following buffer may be made in advance and stored at - 20°C. This buffer will be referred to as buffer "A" throughout the text :

125 mM NaCl
5 mM KCl
1 mM Na$_3$PO$_4$
20 mM HEPES
1 mM MgSO$_4$
1 mg/ml Glucose
0.5 mM CaCl$_2$
pH 7.4 at 37°C

For [Ca^{2+}]i measurements :

a) Buffer A (see above)
b) EGTA (0.5 M, pH 7.4), highest available grade
c) Tris (3 M), highest available grade
d) 10 % Triton X-100
e) 1 M CaCl$_2$

III. EXPERIMENTAL PROCEDURES

A. Preparation of human neutrophils
9.00 am
        (approximately 90 min)
        Freshly drawn blood (plus sodium citrate as anticoagulant and Dextran T500), is allowed to stand at room temperature in a plastic cylinder for approximately 40 minutes, until most red blood cells, RBC, have sedimented. The supernatant is then centrifuged once (1500 rpm's, 10 min) and the pellet containing white blood cells and contaminating RBC is washed once in buffer A and then layered on top of Ficoll Hypaque (usually 7 mls cell suspension are layered over 3 mls Ficoll Hypaque). Cells are centrifuged 20 min 800 x g. Lymphocytes remain at the interphase, while RBC and granulocytes pass through the Ficoll and are found at the bottom of the tube. The pellet is resuspended in a small volume of buffer A (usually 0.5 ml) ; the cells are then subjected to hypotonic lysis to remove RBC by adding 9 mls distilled water, mixing for 20 sec and finally restoring osmolarity by adding 0.5 ml of an 18 % solution of NaCl. The cells (> 95 % neutrophils) are washed, resuspended in buffer A and counted.

B. Loading cells with fura-2 or quin2
10.30 am
        (30 min for fura-2, 60 min for quin2)

        Human neutrophils (approximately 50-100 x 10^6 total) are resuspended in buffer A plus 0.5 % BSA, at a concentration of 5 x 10^7 cells/ml. The cells are pre-incubated in a 37°C waterbath for 5 minutes and then 4 µM fura-2/AM (or 20 µM quin2) diluted in DMSO are added. The suspension is mixed every 2-3 min and after 15 min diluted 5 fold with warm buffer. The incubation is continued for another 15 min (for quin2 loading the total incubation period should be 60 min) ;

finally the cells are washed, resuspended in medium at a concentration of 10^7 cells/ml and kept at room temperature until used. Unloaded cells for autofluorescence should also be prepared ; these should be treated as those receiving dye (i.e. same cell concentration, addition of BSA, incubation at 37°C etc.) but with DMSO alone.

C. Measurements of the intracellular Ca^{2+} concentration (depending on the experiment, approximately 15-20 min per exp.)
12.00 pm

Cells (usually 2.5 x 10^6 cells/ml) should be centrifuged to remove dye which may have leaked out, and resuspended in the fluorimeter cuvette containing warm medium. At the beginning of each experimental day, the fluorescence excitation spectrum (300-400 nm) should be measured, in order to check for complete hydrolysis of the [Ca^{2+}]i indicator. For this purpose the emission peak for quin2 and the excitation peak for fura-2 give the most useful information. After correction for autofluorescence, the emission spectrum of quin2 should peak at $\simeq$ 490-500 nm (339 nm excitation) while for fura-2 the peak in excitation spectrum can vary depending on [Ca^{2+}]i, but should range between 340 and 370 nm. The presence of unhydrolyzed dye is immediately revealed by a shoulder at $\simeq$ 435 nm in the quin2 excitation spectrum and at 380 nm in the emission spectrum of fura-2.

The following is the protocol of a typical experiment with quin2. Neutrophils (2.5 x 10^6 cells/ml) are placed in the fluorimeter cuvette (excitation and emission wavelengths are 339 and 492 nm, respectively) and the pen is set at about 50 % of the total recorder span. Once a steady level of fluorescence is reached (quin2 and fura-2 fluorescence is temperature sensitive, higher at lower temperature), 10^{-7} M fMLP, is added to the cuvette. This results in an immediate increase in fluorescence which peaks at variable times (depending on the intracellular quin2 concentration), and returns to basal levels again depending on the intracellular concentration of the dye and the presence or absence of extracellular Ca^{2+}. Once fluorescence has returned to or near basal, a calibration to calculate $[Ca^{2+}]i$ is performed ; below we will describe one of the several possible calibration procedures. First EGTA (5 mM final concentration) and Tris (30 mM) are added ; the downward jump in fluorescence is taken as a measure of external quin2. The cells are then lysed with 0.025 % Triton-X 100. This value of fluorescence is called Fmin. Finally Ca^{2+} is added back to the cuvette, enough to compensate the EGTA. This new value is referred to as Fmax. The exact same experiment is performed with unloaded cells in order to check for changes in cellular autofluorescence.

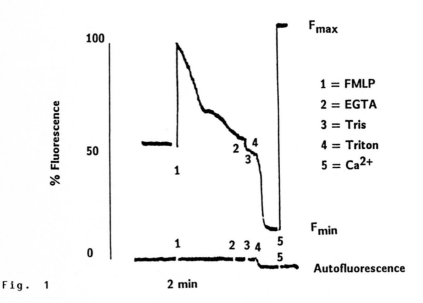

Fig. 1   2 min

D. Calculating [Ca^{2+}]i

The first step is to correct for changes in autofluorescence, if they occur. In the example illustrated in Fig. 1,
the only compound affecting autofluorescence is Triton X-100
(a downward deflection) ; this should be corrected by increasing the Fmin and Fmax values by the same amount. These values
are then corrected for the dilution, due to the various
additions. These new values are called Fmin* and Fmax*. Next
one must calculate the contribution of external quin2, i.e.
the downward jump after EGTA. The contribution of extracellular dye to the fluorescence F (for example the resting value)
is then substracted (to become the new value F*) and [Ca2+]i
can be calculated from the formula :

$$(F* - Fmin*)/F(max\% - F) \times 115 \text{ nm}$$

IV. REFERENCES

Andersson T., Dahlgren C., Pozzan T., Lew D. (1987). Characterization of fMet-Leu-Phe receptor mediated Ca^{2+} influx
across the plasma membrane of human neutrophils. **Mol. Pharmacol.**, **30** : 437-443.

Berridge M.J. (1984). Inositol trisphosphate and diacylglycerol as second messengers. **Biochem. J.**, **220** : 345-360.

Berridge M.J., Irvine R.K. (1984). Inositol trisphosphate, a
novel second messenger in cellular signal transduction. **Nature, 312** : 315-321.

Campbell A.K. (1983). Intracellular Ca^{2+} its universal role as
a regulator. John Wiley and Sons, Chichester.

Di Virgilio F., Vicentini L.M., Treves S., Riz G., Pozzan T.
(1985). Inositol phosphate formation in fMet-Leu-Phe stimulated human neutrophils does not require an increase
in the cytosolic free Ca^{2+} concentration. **Biochem. J.**,
229 : 361-367.

Grinkiewicz G., Poenie M., Tsien R.Y. (1985). A new generation
of Ca^{2+} indicators with greatly improved fluorescence properties. **J. Biol. Chem.**, **260** : 3440-3450.

Nishizuka Y. (1984). The role of protein kinase C in cell
surface signal transduction and tumor promotion. **Nature,**
308 : 693-698.

Experiment n° 10

SYNTHESIS AND MATURATION OF D-β-HYDROXYBUTYRATE DEHYDROGENASE (BDH) FROM MITOCHONDRIAL INNER MEMBRANE

A. KANTE, J.M. BERREZ & N. LATRUFFE

I. INTRODUCTION AND AIMS

Mitochondria contain several hundred proteins. Most of them are encoded by nuclear DNA, synthetized in the cytosol and post-translationnally imported into mitochondria (see review of Douglas et al., 1986 ; Hurt & Van Loon, 1986).

How are transported cytoplasmically made proteins into the mitochondria ? This question, one of the central problems of mitochondrial biogenesis, can be approached by techniques like **in vitro** translation and **in vitro** processing.

Indeed mitochondria can take up post-translationnaly the **in vitro** neosynthetized mitochondrial precursors and translocate them into their proper internal compartments by an energy dependant process (Miralles et al., 1982). This import is accompagnied by a proteolytic maturation of precursors exhibiting a presequence. Previous work has confirmed that the import of polypeptides **in vitro** reflects the process as it occurs in living cells. The evidence of the import is proved by the fact that up to 70 % of a given precursor synthetized **in vitro** can be processed by mitochondria and protected from proteases added to the incubation mixture (Gasser et al., 1982). It is noteworthy that the imported polypeptides are localized in their correct submitochondrial compartment (Ono and Tuboi, 1986).

The uptake requires both an electrochemical potential across the mitochondrial inner membrane and triphosphate nucleotides (Pfanner & Neupert, 1986).

Immunoprecipitation has become a widespread technique for the identification of low abundance **in vitro** translated and processed products (less than 1 % of the total synthetized protein). If you associate immunoprecipitation, PAGE-SDS and fluorography, then, you can distinguish the precursor (if it contains a signal presequence) from the mature form (processed in the mitochondria by a matrix protease).

Our model will be the D-β-hydroxybutyrate dehydrogena-se from rat liver, an inner membrane lipid requiring enzyme (see review Latruffe, 1986) involved in the metabolism of the ketone bodies (Lehninger et al., 1960). This protein is synthetized on free cytosolic polysomes as a larger precursor and processed into mitochondria (Kanté et al., 1987).

This method can be exactly repeated for any kind of cytoplasmically made mitochondrial protein or could be extended to study the insertion or translocation of newly synthetized protein into organels.

The experiment will be processed in three steps :

a. **In vitro** translation in a reticulocyte lysate programmed with free polysomes from rat liver.

b. Processing of precursor by freshly isolated mitochondria.

c. Immunoprecipitation followed by electrophoresis and fluorography.

II. EQUIPEMENT, CHEMICALS AND SOLUTIONS

A. Equipment

1. Access to

- A refrigerated ultracentrifuge Beckman (Type 40 rotor)
- A preparative centrifuge Sorvall (SS 34 rotor)
- A low speed centrifuge (Jouan) with conical plastic tubes
- A microcentrifuge for Eppendorf tubes
- A liquid scintillation counter with vials and scintillation liquid
- A spectrophotometer (595 nm for Bradford protein assay)
- An oxygraph
- A potter-Thomas homogenizer with a motor- driven Teflon pestle
- A "carrousel" for tubes agitation in the cold room
- A vertical slab gel electrophoresis apparatus with a power supply
- A gel dryer (LKB)
- A water vacuum pump

2. On the bench

- A thermostated water bath with rack for Eppendorf tubes
- A dissection board
- A vortex
- A pH meter
- A magnetic stirrer with Teflon-coated magnetic rods
- Sterile gloves
- Sterile tips, sterile Pasteur pipettes and sterile Eppendorf tubes
- Glassware, automatic pipettes (5 µl to 5 ml) with tips, and disposable plastic tubes
- A timer

- 3 MM Whatman paper, benchcoat
- A Hammilton syringe (100 µl)

Caution : The basic biological material for in vitro protein synthesis is polysomes i.e. RNAs. So it is very important to protect them against RNAses. The skin is a major source of RNAse. To avoid RNAse contamination you must use gloves, autoclaved tubes and tips. The glassware must be acid clean and baked (4 hours at 180°C).

B. Chemicals and biological material

- A starved rat
- Rabbit reticulocyte lysate (RRL) stored at - 196°C (from Amersham or prepared according to Pelham & Jackson 1976)
- Rat liver cytosolic free polysomes stored at - 80°C, prepared according to Ramsey & Dawley, 1976
- Polyclonal and specific rabbit antiserum against rat liver BDH, stored at - 20°C
- Ultrapure and sterile water from MilliQ (Millipore)

C. Solutions (RNAse-free, can be stored for months at - 20°C)

a) In vitro protein synthesis

- Micrococcal nuclease (1 mg/ml in 10 mM K.Hepes pH 7.5) from Boehringer
- Hemin-HCl 1 mM from Calbiochem
  (6.5 mg hemin-HCl, firstly dissolved in 0.125 ml 2 M KOH + 0.625 ml H_2O + 0.100 ml 2 M Tris-HCl + 8.9 ml ethyleneglycol, final pH = 7.0)
- 0.1 M $CaCl_2$
- 0.1 M EGTA pH 7.0

- Master Mix

  (0.2 mM aminoacid mixture **minus** methionine ; 2 mM GTP ; 5 mM
  ATP ; 0.4 M K Acetate pH 7.5 ; 10 mM Mg Acetate ; 75 mM K
  Hepes pH 7.5)
- 13.7 mg/ml tRNAs from calf liver from Boehringer
- ^{35}S Methionine (> 1000 Ci/mmole) ref. SJ204 from Amersham
- RNAsin from Amersham (facultative)
- 100 mM creatine phosphate pH 7.0 from Boehringer
- 10 mg/ml creatine phosphokinase from Boehringer
- 10 mM spermidine from Boehringer

b) Import of the precursors into mitochondria (Kuzela et al., 1985)

- BMB : 50 mM Bicine ; 5 mM KH_2PO_4 ; 90 mM KCl, pH 7.6 with
  5 M KOH
- Sephadex G-25 (Medium) swollen and many times decanted in
  BMB
- RLM buffer (for mitochondria preparation) : 0.25 M sucrose ;
  10 mM K-Hepes ; 1 mM K-EDTA, pH 7.4
- 50 mM $MgCl_2$
- 0.1 M ATP pH 7.0 (adjust with KOH)
- 0.1 M Phosphoenolpyruvate pH 7.0 adjust with KOH)
- Pyruvate kinase 10 mg/ml from Boehringer
- 0.5 M DTT from Boehringer
- Ernster medium containing 100 mM KCl, 5 mM $MgCl_2$, 50 mM Tris
  HCl, pH 7.5
- 1 M K Pi pH 7.0
- 1 M succinate pH 7.0
- 1 M malate + glutamate pH 7.0
- 50 mM ADP pH 7.0

c) Immunoprecipitation of BDH and its precursor

- 20 % SDS (electrophoresis grade, from Serva)
- TNET : 50 mM Tris-HCl ; 0.3 M NaCl ; 5 mM Na-EDTA ; 1 %
  Triton X 100, pH 8.0
- TNE : 50 mM Tris-HCl ; 0.15 M NaCl ; 5 mM Na-EDTA, pH 7.5
- TE : 50 mM Tris-HCl ; 5 mM Na-EDTA, pH 7.5
- Trasylol (aprotinin) : 40 000 Kalikrien units/ml of 10 mM K
  Hepes pH 7.4 from Boehringer
- Facultatively coktail of protease inhibitors (1:1:1) :
    * TPCK 0.2 M in DMSO from Sigma ⎫
    * TLCK 0.2 M in DMSO from Sigma ⎬ can be stored at - 20°C
    * PMSF 0.2 M in ETOH from Sigma ⎭
- Protein A-Sepharose (P.A. Sepharose) from Pharmacia
- Extraction buffer : 110 mg/ml SDS ; 15 mM Tris-HCl pH 6.8 ;
  30 % glycerol

III. EXPERIMENTAL PROCEDURES

a) In vitro protein synthesis

Day 1
8.45 am

1. Nuclease treatment

Purpose : to destroy endogenous rabbit mRNAs in the reticulocyte lysate, unless you get commercial nuclease treated reticulocytes (Amersham).

- Take a tube of lysate (500 μl) from liquid nitrogen
- Lay 13 μl of hemin on the top of the tube frozen lysate. Then thaw it with hand while mixing with Pasteur pipette to rapidly homogenize hemin (avoid foaming)
- Withdraw 19 μl and keep it on ice

- Mix (thoroughly, avoid foaming)
  * 494 µl lysate
  * 4.8 µl 100 mM CaCl$_2$
  * 2.4 µl micrococcale nuclease
and incubate 8 mn in water bath at 20°C.

- Stop the reaction by adding 14.3 µl of 0.1 M EGTA, mix thoroughly and put it on ice.

2. Protein synthesis

- In a tube prepare synthesis mixture (SM) :

210   µl master mix
50   µl creatine phosphate
20   µl creatine   phosphokinase
16   µl spermidine
14.5 µl tRNAs

_____

=   310.5 µl synthesis mixture

- Mix 500 µl of nuclease-treated lysate with 290 µl SM = 790 µl (RRL + SM) for samples 2 and 3.

- Mix 19 µl of non nuclease-treated lysate with 11 µl SM for sample 1.

- Prepare your different samples in Eppendorf tubes as follows :

Sample	Mixture RRL + SM	RNAse treatment	Polysomes (40 OD_{260}/ml)	^{35}S-Met (1 mCi/ml)
1	30 µl	no	-	30 µCi
2	30 µl	yes	-	30 µCi
3	395 µl	yes	x µl	y µCi

- Add radioactive methionine in last, mix thoroughly but gently and incubate 45 mn at 30°C in a water bath.

- At different times (0, 5, 10, 15, 30 and 45 mn) withdraw 2 µl of labeled lysate and spot it on a numbered 3 MM Whatman paper square (1 cm²).

- After 45 mn, centrifuge the sample n° 3 during 45 mn at 105 000 g (rotor type 40) and discard the polysomal pellet. Withdraw 2 µl of the supernatant and spot it on Whatman paper.

It is possible to freeze this labeled lysate and to store it at - 80°C for further use.

3. Translation efficiency
10.00 am

Incorporation of ^{35}S-Methionine into newly synthetized proteins is determined by counting the aliquots applied to Whatman squares, which have been processed for hot trichloro-acetic acid-insoluble radioactivity estimation.

Procedure :

    - Transfer the dry Whatman papers in a beaker containing 7 % TCA for 20 mn

    - Transfer the samples in a boiling 7 % TCA solution containing 1 mg/ml of non-labeled methionine. Leave for 15 min on the heating plate.

    - Rince the Whatman papers with 7 % TCA

    - Rince for a few minutes with 75 % EtOH 25 % Ether (V/V)

    - Rince with ether

    - Dry the Whatman papers with a hair dryer

    - Transfer them into scintillation vials with 5 ml of scintillator liquid and count the samples for 1 min.

Results :

    The first and the second samples are used to estimate the nuclease treatment efficiency.

    Typical results are :

    - 50 000 cpm per μl for non nuclease-treated lysate without rat polysomes

    - 5 000 cpm per μl for nuclease-treated lysate without rat polysomes

    - between 250 000 and 500 000 cpm per μl for nuclease-treated lysate with 40 $OD_{260\ nm}$/ml of rat liver polysomes

    - between 150 000 and 300 000 cpm per μl after polysome sedimentation by ultracentrifugation.

b) Import of the precursors into mitochondria (Kuzela et al., 1985)

    In order to improve the import efficiency, you can exchange acetate ions with chloride ions using the gel filtration-centrifugation method :

The translated lysate is chilled to 4°C and filtered
through a Sephadex G-25 column, prepared in a 1-ml disposable
syringe, and equilibrated with BMB.

To avoid dilution of the lysate, the excess of
equilibration buffer is removed by centrifugation (1500 rpm, 1
mn in medical centrifuge). The lysate is then applied to the
column, and the centrifugation is repeated. Filtered lysate is
collected in an Eppendorf tube set to the syringe bottom (see
Figure 1). This step removes the excess of free ^{35}S-Methioni-
ne, metal ions and other small molecules from the translated
lysate.

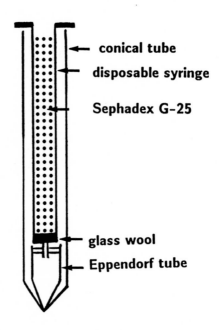

Figure 1

10.00 am

1. Rat liver mitochondria preparation

- A single liver is quickly removed from a starved animal and placed in ice-cold RLM buffer.
- Cut it in pieces and rince it to remove blood (until medium is clear).
- Homogenize it with a large potter with 40 ml of RLM buffer (with motor at 220 rpm - 3 up and downs).
- Centrifuge the homogenate in Sorvall centrifuge : 10 mn at 2500 rpm (rotor SS 34).
- Centrifuge the supernatant : 15 mn at 8000 rpm (rotor SS 34).
- Discard the supernatant and then clear the tube walls with Kleenex to remove fat.
- Resuspend the mitochondrial pellet with a 2 ml-glass pipette in 20 ml of buffer (avoid bubbles and rough manipulations).
- Centrifuge 10 mn at 9000 rpm (rotor SS 34).
- Resuspend the pellet in 10 ml buffer.
- Repeat the washing in 5 ml buffer.
- Finally resuspend the mitochondria in a minimal volume of RLM buffer (less than 1 ml).
- Evaluate the protein concentration with biuret 15 mn (Jacobs et al., 1956) or Bradford 5 mn (Bradford, 1976) techniques.
- Make sure that mitochondria are not damaged (still coupled) in respiratory control assay (Chance & Williams, 1956).

1.30 p.m.

2. Oxygraphy study

- Preheat 1.8 ml of modified Ernster medium in the oxygraph cell (30°C)

- Add 2 mg mitochondrial proteins and 10 µl K Pi
- Wait until the base line is stable
- Add 10 µl of succinate or malate + glutamate
- Wait until the respiration line slope is stable (state 4)
- Add 5 µl of ADP, record the state 3
- Determine the respiratory control ratio (RCR = state 3/state 4).

N.B. : The RCR must be at least 3 to proceed further.

2.00 pm

3. Processing experiment

The sample must contain at least 10-20 10^6 cpm of methionine incorporated in total proteins.

All operations are performed at 0°C and the uptake is started by transferring the tubes (preferably Eppendorf tubes) into a 27°C waterbath.

- Add previously centrifuged (105 000 g during 45 min) non-labeled lysate to the final concentration of 20-30 mg/ml.

N.B. : If you have previously gel filtered the labeled lysate, this procedure restores the non-mitochondrial factors needed for import.

- Add :

   * $MgCl_2$ to final concentration of 1 mM
   * ATP to final concentration of 1 mM
   * PEP to final concentration of 1 mM

        * Pyruvate kinase to final concentration of 0.1 mg/ml

        * DTT to final concentration of 1 mM

- Add freshly isolated mitochondria to the final concentration of 2 mg/ml of proteins.

- Gently mix and incubate at 27°C.

- After 1 hour incubation, centrifuge the suspension in a Eppendorf micro-centrifuge (5 mn in cold room) to separate supernatant from mitochondrial pellet.

- Overlay the pellet with RLM buffer in order to do a rapid rincing of the surface.

- Resuspend the pellet in RLM buffer (in the same volume than the supernatant).

c. Immunoprecipitation of BDH and its precursor

1. Suspension of P.A. Sepharose in TNET

Can be prepared from the morning as follows :

Swell the Sepharose beads in TNET during several hours. Wash the swollen beads several times with TNET by resuspension and a short and low centrifugation (time just required to raise up to 3000 rpm).

Add 1 volume of TNET to 1 volume of packed P.A. Sepharose. Keep it at + 4°C.

4.30 pm

2. Immunoprecipitation

- Add 0.2 volume of 20 % SDS to the samples (mitochondria and supernatant).
- Boil for 3 mn the samples in a waterbath.
- Immediately transfer the samples in conical disposable tubes (10 ml) and dilute them with 15 volumes of cold TNET.
- Add the protease inhibitors : Trasylol (5 µl/ml) and facultatively TPCK : TLCK : PMSF mixture (15 µl/ml).
- Add clarified anti-BDH serum : 15 µl serum/supernatant sample or 30 µl serum/mitochondrial sample (which contains a lot of endogenous BDH).
- Incubate overnight in the cold room on the carrousel.

Day 2

7.00 am

- After the overnight incubation with serum, P.A. Sepharose is added to the samples : (minimum 2 volumes of 1:1 Sepharose P.A./volume of serum). We shall use 100 µl.

- Incubate 2 hours in the cold room on the carrousel.

9.00 am

- Spin 2 mn in a medical centrifuge (Jouan) at 3000 rpm to collect the P.A. Sepharose beads. You can use the supernatant to immunoprecipitate other proteins.

- Wash the P.A. Sepharose containing the immune complexes :

* 5 times with TNET
* 3 times with TNE
* 1 time with TE

Fill up the tubes to the top during washing.

- Transfer completely the beads into Eppendorf tubes using TE for rincing.
    - Spin 2 min and discard the TE supernatant.
    - The separation of the immune complexes from the Sepharose beads is performed by adding 90 µl of extraction buffer and heating 3 mn at 100°C in a waterbath.
    - Centrifuge 3 mn and collect the supernatants.
- Withdraw aliquots (10 µl) for counting and apply the rest on a electrophoresis gel.

3. <u>Electrophoresis according to Laemmli (1970)</u>

For mature BDH (MW = 31.5 KD) and preBDH (34.5 KD), we use 12.5 % polyacrylamide-SDS gel.

Separation gel (12,5 %) :
* 15 ml 30 % acrylamide-bisacrylamide
* 11.5 ml H_2O
* 9 ml Lower Buffer (1.5 M Tris, HCl pH 8.8)
* 0.36 ml 10 % SDS
* 0.15 ml 10 % ammonium persulfate (extemporally prepared)
* 15 µl TEMED

Stacking gel (3 %) :
* 1 ml 30 % acrylamide-bisacrylamide
* 6.3 ml H_2O
* 2.5 ml Upper Buffer (0.5 M Tris-HCl pH 6.8)
* 0.1 ml 10 % SDS

    * 0.1 ml 10 % ammonium persulfate
    * 10 µl TEMED
Running buffer :
    * 25 mM Tris
    * 192 mM Glycine   }   pH 8,3
    * 0.1 % SDS

    Samples : Add 2 µl 0.025 % Bromophenol Blue and 2 µl
β-mercaptoethanol to the extracted samples.

    Don't forget to load molecular weight markers and
purified native BDH and to fill up the empty pockets with the
extraction buffer.

11.00 am
    Run 5 hours at 30 mA in the cold room.

4.30 pm
    4. Fluorography

    Stain the gel with Coomassie Blue solution (0.25 % in
10 % methanol, 7 % acetic acid and destain it with 50 %
methanol ; 7 % acetic acid).

6.30 pm
    Impregnate the gel with 1 M Na-salicylate from Merck
(150 ml/gel 20 mn shaking).

    Wash quickly but extensively the gel with distilled
water to avoid further salicylate precipitation.

    Dry the gel for 2 hours on a sheet of Whatmann paper
with the gel-dryer.

8.30 pm

The dried gel is then exposed for fluorography on Amersham Hyperfilm or X-0 mat Kodak film and developped after the appropriate period. For instance 10 days exposure are required for 1000 cpm.

Typical results :

24.10^6 cpm before processing experiment

2400 cpm immunoprecipited BDH in the lysate (0.01 %)

800 cpm incorporated BDH in the mitochondria (0.0033 %)

Figure 2 : Fluorography film

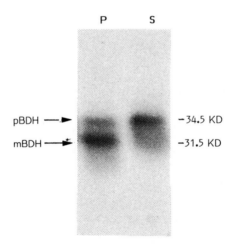

pBDH = precursor form of BDH
mBDH = mature form of BDH
P = mitochondrial pellet
S = supernatant

IV. COMMENTS

The fluorography film clearly shows molecular weight difference between native BDH (or processed BDH in mitochondria) and precursor. The leader sequence size can be estimated at 3 KD. It seems to be removed in one step (no intermediate precursor has been revealed in the mitochondria). The two forms (precursor and mature) observed in the mitochondria indicates two events in the incorporation process :

1 - binding of the precursor to the outer membrane.

2 - translocation across the outer and inner membranes and processing by the matrix protease.

Troubleshooting :

Lack of specific immunoprecipitation : If you don't succeed to immunoprecipitate a peculiar **in vitro** synthetized protein, numerous reasons have to be considered. Some are obvious, such as degradation of mRNA, poor radioactive incorporation into protein or proteolysis. With membrane proteins however, there is an additional possibility that the antiserum may not react with the primary translation product. This may be especially true if the precursor and the mature protein do not exhibit the same conformational antigenic determinants (posttranslational modifications). The way to minimize this possibility is to prepare the original antibody against SDS-denatured antigen.

V. REFERENCES

Bradford M.M. (1976). A rapid and sensitive method for the quantitation of microgram quantitie of protein utilizing the principle of protein-Dye binding. **Anal. Biochem., 72** : 248-254.

Chance B. (1959). Quantitative aspects of the control of oxygen utilization in **Ciba foundation symposium on regulation of cell metabolism,** pp 91-129, Churchill, London.

Douglas M.G., Mc Cammon M.T. & Vassaroti A. (1986). Targeting proteins into mitochondria. **Microbiol. Rev., 50** : 166-178.

Gasser S.M., Daum G. & Schatz G. (1982). Import of proteins into mitochondria. **J. Biol. Chem., 257** : 13034-13041.

Hurt E.C. & Van Loon A.P.G.M. (1986). How proteins find mitochondria and intramitochondrial compartments. **Trends Biochem.Sc., 11** : 204-207.

Jacobs E.E., Jacob M., Sanadi D.R. & Bradley L.B. (1956). Uncoupling of oxidative phosphorylation by cadmium ion. **J. Biol. Chem., 223** : 147-156.

Kanté A., Latruffe N., Duroc R. and Nelson B.D. (1987). Import and processing of D-β-hydroxybutyrate dehydrogenase larger precursor into mitochondria after **in vitro** translation of rat liver free polysomes. **Life Sci. Adv., 6** : 121-123.

Kuzela S., Joste V. & Nelson B.D. (1986). Rhodamine 6G inhibits the matrix-catalysed processing of precursors of rat liver mitochondrial. **Eur. J. Biochem., 154** : 553-557.

Laemmli U.K. (1970). Cleavage of structural proteins during the assembly of the head of bacteriophage T4. **Nature, 227** : 680-685.

Latruffe N., Berrez J.M. and El Kebbaj M.S. (1986). Lipid-protein interactions in biomembranes studied through the phospholipid specificity of D-β-hydroxybutyrate dehydrogenase. **Biochimie, 68** : 481-491.

Lehninger A.L., Sudduth H.C. and Wise J.B. (1960). D(-)-β-
hydroxybutyric dehydrogenase of mitochondria. J. Biol.
Chem., 235 : 2540-2545.

Macchecini M.L, Rudin Y., Blobel G. and Schatz G. (1979).
Import of protein into mitochondria. Precursor forms of the
extramitochondrialy made F1-ATPase subunits in yeast. Proc.
Natl. Acad. Sci. (USA), 76 : 342-347.

Miralles V.J., Felipo V., Hernandez-Yago Y. (1984). Synthesis
and transport of the precursor for the β-subunit of rat
liver F1-ATPase. Physiol. Chem. and Phys. and Med. NMR,
16 : 513-519.

Ono H. & Tuboi S. (1986). Translocation of proteins into rat
liver mitochondria. The precursor polypeptides of a large
subunit of succinate dehydrogenase and ornithine amino-
trans-
ferase and their import into their own locations of mitochondria. Eur. J. Biochem., 155 : 543-549.

Pelham H.R.B. & Jackson R.J. (1976). An efficient mRNA-depen-
dent translation system from reticulocyte lysates. Eur. J.
Biochem., 67 : 247-256.

Pfanner N. & Neupert W. (1986). Transport of F1-ATPase subunit
β into mitochondria depends on both a membrane potential
and nucleoside triphosphates. FEBS lett., 209 : 152-156.

Ramsey J.C. & Steele W.J. (1976). A procedure for the quantitative recovery of homogeneous populations of undegraded
free and bound polysomes from rat liver. Biochemistry, 15 :
606-615.

Experiment n° 11

ELECTROPERMEABILIZATION AND ELECTROFUSION OF CELLS

J. TEISSIE & M.P. ROLS

- More than a way to <u>zap</u> a cell -

(Biotechnology (1985) 3 187)

I. INTRODUCTION AND AIMS

Cell electropermeabilization is a very powerful tool for cell biology. When performed under suitable conditions, the electric field induction of a new organization of the plasma membrane leads it to a "permeabilized" state. This organization is transient but can nevertheless be maintained for some minutes. The cytoplasmic content can then be changed without any harmful effect on the cell viability by simply incubating the cells after electric pulsation in a new buffer containing the compounds of interest. Molecules with molecular weight up to 1 kDa and nucleic acids can then cross the plasma membrane. An equilibration between the outer and inner cellular volumes occurs during the transient permeable state. With the spontaneous reversibility of the membrane organization, the cell recovers a normal state excepted that the content of the cytoplasm is the one we have choosen.

Due to the vectorial character of the interaction between field and cell (Teissié & Blangero, 1984), the new organization affects only localized patches of the cell membrane, facing the electrodes (Sowers, 1986). It is triggered only when the field intensity is larger than a given threshold (a function of the cell strain) and is under the control of the pulse duration (the longer the pulse, the larger the molecules which can permeate across the plasma membrane (Schwister & Deuticke, 1985). In conclusion, the electropermeabilization is under the control of :

    a) the field intensity
    b) the pulse duration
    c) the number of pulses
    d) the direction of the field.

The electropermeabilization is used for the introduction of small molecules inside the cytoplasm under viable conditions (Knight & Scrutton, 1986), for the transformation of mammalian cells (Neumann et al., 1982 ; Zerbib et al., 1985) or of plant protoplasts (Shillito, 1985). The other advantage of the electropermeabilization is the induction of a "fusogenic" character of the cell surface (Sowers, 1986 ; Teissié & Rols, 1986). High yields of viable hybrids can thus be obtained.

Reviews on the subject have been published (Zimmermann, 1982 ; Neumann, 1984) but the state of knowledge on the subject is in constant progress.

II. EQUIPMENT, CHEMICALS AND SOLUTIONS

A. Equipment

1. Description of the electropulsator

The electropulsator consists of the following units :

1. A high power electric generator, which is the slave of :

2. A microcomputerized clock

3. A display oscilloscope, where the intensity and the duration of the electric pulses are visualized on line.

4. A set of electrodes which are in contact with the culture dish.

In order to select the field intensity E, the width between the electrodes, d is checked and the voltage V which is going to be delivered by the generator is set at :

$$V_{(kV)} = E_{(kV/cm)} * d_{(cm)}$$

(check carefully your units !)

Then select the pulse duration T (in the microsecond range) and the delay between the pulses (this delay is the reciprocal of the frequency of the pulses). You can check the duration on the scope (channel B). The field pulse can then be triggered either by :

1. The manual way : the pulse is applied when you push the switch
2. The automatic way : the successive pulses are applied in a row when you push the switch, you stop it when the selected number of pulses has been applied.

The pulses are square wave. This means that the field intensity is constant during the pulse duration even when the ionic content of the buffer is high (up to 0.15 M). This is controled on line by means of the scope. If something is wrong, erratic signals are observed, warning you at once (check your buffer !).

The major advantage of the "square wave" pulse technique is that the external field intensity remains constant during the pulsation allowing an accurate control of the perturbations that you applied on the cell culture. You can select a field intensity which is suitable for permeabili-zation with a pulse duration which does not alter the cell viability (duration as long as 150 μs) and this can be obtained whatever the content of your pulsing buffer. This is not the case with the capacitor system (exponentially decaying

field) where the field intensity is continuously decreasing
with time and where the characteristic decay time is a
function of the pulsing buffer. Furthermore, in most experiments, the high field intensity which is present at the
beginning of the field decay, is toxic for the cells, but
nevertheless is needed to obtain a good permeabilization. In
contrast, the optimization of the electropermeabilitzation
(i.e. 100 % of viable and permeabilized cells) is possible
routinely with the square wave technique.

Sterility of the electrodes is obtained by washing
them with ethanol and by rinsing carefully with sterile water.

2. Cell culture

- An humidified CO_2 incubator
- An incubator
- A laminar flow hood
- An ice bucket
- An inverted microscope with phase contrast optics
- Adams Tally counters
- Culture dishes with a diameter of 35 mm (Nunc is our first
  choice)
- Culture flasks with a surface of 25 and 80 cm² (Nunc is our
  first choice)
- Pasteur pipets
- Digital pipets (Pipetman P200 and P1000, Gilson or equiva-
  lent) with cones (yellow and blue)
- Single use sterile pipets (1 ml and 10 ml)
- Millex filters with a pore diameter of 0.22 µm (Millipore or
  equivalent)
- Single use syringes (10 ml)

B. Material

1. Chemicals

- Eagle's medium MEM
- Dulbecco's MEM
- Serum

2. Miscellaneous

- Gloves
- Timer
- Trypan blue
- Movable stage (Boy)
- Phosphate buffer (pH = 7.4)
- Salts and sugar (analytical grade)

3. ATP determination

- Luciferin-Luciferase kit (Sigma L 0633)
- Glass scintillation vials
- Scintillation counter
- Glycylglycin buffer (pH = 7.4)

4. Electropulsator

The generator is described in II.A (patent CNRS).

3. Some experimental details

The volume of pulsing buffer which is added is 1.5 ml.

Cells growing on a culture flask (25 cm^2) are treated during 30 s. at 21°C by a mixture of 1 ml of Trypsin (0.1 % w/v) and 1 ml of EDTA (0.3 % w/v), both in Ca^{++} Mg^{++} free PBS, and the cells are then incubated at 37°C during 5 min. after removal of the solution. The cells are then washed with the culture medium containing the serum to stop the effect of Trypsin. The cells are detached from the flask surface by shaking and are dispersed in suspension in the culture medium.

The cell density is set at 1.5 10^5 cells/ml for permeabilization experiments and to 4 10^5 cells/ml for fusion experiments. 2 ml of cell suspension are added per culture dish. The cell in the dish must be kept for more than 2 hours in the CO_2 incubator in order to get plated.

C. Solutions

1. For cell culture

Cells (Chinese hamster ovary - CHO - or mouse fibroblasts - 3T3 -) are grown on culture dishes (diameter of 35 mm) in a CO_2 incubator (5 %, 37°C) in culture medium (MEM or DMEM) with 6 to 10 % serum, antibiotics and glutamin.

The culture should be non confluent for the permeabilization experiments and confluent for the fusion studies.

2. Electropulsation buffers

A -   Phosphate buffer 10 mM pH = 7.4
       $MgCl_2$ 1 mM
       Sucrose 250 mM (for osmolarity)

B -   Phosphate buffer 10 mM pH = 7.4
       $MgCl_2$ 1 mM
       NaCl 100 mM
       Sucrose 84 mM

C -   Phosphate buffer 10 mM pH = 7.4

$MgCl_2$ 1 mM

KCl 100 mM

Sucrose 84 mM

(Other pH buffers - Tris, MES... - can be used in place of the phosphate buffer ; the counter ion can be Na^+ or K^+).

3. Staining buffers

Trypan blue (4 mg/ml) is added to the pulsing buffer.

D -   A plus TB

E -   B plus TB

F -   C plus TB

4. Saline buffer

Phosphate buffer 10 mM pH = 7.4

$MgCl_2$ 1 mM

NaCl 150 mM

III. EXPERIMENTAL PROCEDURES

A. Electropermeabilization
8.45 am $\longrightarrow$ 11.45 am

Cell electropermeabilization allows the penetration of small molecules (with molecular weight up to 1000 Da) inside the cytoplasm. The process is easily observed by use of non membrane permeant nuclear stains such as Ethidium bromide or Trypan blue. This last one is used in the present experiment because a transmission microscope is just required. Permeabilized cells have their nuclei stained in deep blue. Our technical approach where cells are pulsed directly on the

culture dish gives an "internal" control. Cells which are
between the electrodes are the only ones which are pulsed ;
cells which are outside the width between the electrodes are
not submitted to the field and are control cells.

The purpose of the study is to determine the experimental parameters (field intensity, pulse duration and number
of pulses) that affect electropermeabilization. The delay
between the pulses (up to 30 s.) has been shown not to play a
role. Optimization of electrofusion (100 % of permeabilized
viable cells) requires the control of these parameters for a
given cell strain but depends on the cell strain and for a
given strain whether the cells are plated or in suspension
(after trypsinization).

Experimental procedure

1. Non confluent cells are grown on the culture dish.

2. The culture medium is carefully removed by use of a
Pasteur Pipet.

3. 1.5 ml of the staining buffer (D, E or F) is added.

4. The electrodes are brought into contact with the
bottom of the dish by means of the movable stage.

5. Trigger the pulses after selection of the parameters (E, T, n) (begin with 5 pulses of 100 μs and field
intensity of 0.1 kV/cm).

6. During 5 check on the scope the shape of your
pulses (you should observe a square wave).

7. Move down the dish by the movable stage, select a
new place on the culture which has not been between the
electrodes.

8. Bring the electrodes in contact with the bottom of the dish as in 4.

9. Trigger the pulses as in 5 but with a new set of parameters (if 5 was as described then use 5 pulses of 100 µs and a field intensity of 0.2 kV/cm).

10. If it is possible do again step 7.

11. The cells are incubated during 5 min. at room temperature.

12. Pipet out the staining buffer and wash with the saline buffer G.

13. Observe the cell culture under the microscope (no color filter, the phase ring can be omitted). Use objectives of 20x or 32 x.

14. Observe different areas on the dish.

15. Count in the different pulsed places and in the control areas the cells where the nuclei is blue and the total number of cells (you must observe at least 250 cells per experimental condition for statistical reasons). Stained cells are either dead (their percentage of the population is the one of the control) or permeabilized.

16. Be back to step 2 with a new culture dish and new pulse parameters in 5.

17. Plot the change in percentage of permeabilized cells as a function of the parameter (E, T or n).

E (kV/cm)	0.1	0.2	0.3	0.4	0.5	0.6	0.7	0.8	0.9	1	1.2	...
P (%)												

T (μs)	5	10	20	...
P (%)				

18. Does the composition of the pulsing buffer play a role in electropermeabilization ?

B. Reversibility of electropermeabilization
1.15 pm ⟶ 3.00 pm

The electric field induced permeabilization is a reversible process. The new organization of the membrane dissappears spontaneously as a consequence of a metabolic reaction of the cell to the "stress". This reversibility is under a strict control of the temperature.

This reversibility is easily observed by adding the staining buffer after various delays following the pulses. With increasing delays, less and less cells are stained.

Experimental procedure

1. Non confluent cells are grown on a culture dish.

2. Remove the culture medium with a Pasteur pipet.

3. Add 1.5 ml of the pulsing buffer (A, B or C).

4. Bring the electrodes into contact with the bottom of the dish by use of the movable stage.

5. Trigger the electric pulses with the suitable set of parameters (5 pulses of 1 kV/cm and a duration of 100 µs is a good choice for plated cells).

6. Check during 5 the shape of the pulse on the scope.

7. Switch on the timer.

8. Move down the culture dish by use of the movable stage.

9. Incubate the culture dish at the right temperature (if you are using 4°C, a good idea is to have it already on the ice bucket during the pulses).

10. After a selected delay (at least 15 s.), remove the pulsing buffer with a Pasteur pipet and add the associated staining buffer (1.5 ml).

11. Incubate 5 min. at room temperature.

12. Remove the staining buffer and wash with the saline buffer G.

13. Observe the cell culture with an inverted micros- cope (no color filter, the phase ring is not needed, objective 20x or 32x).

14. Determine the percentage of the cell population which is blue stained in the width between the electrodes and in the control area (more than 250 cells must be observed in both cases).

15. Do again the experiment with another culture and the same set of parameters but another delay between the pulses (step 5) and the staining (step 10).

16. Conclusion on the dependence of reversibility on time, on temperature.

C. Leak of cytoplasmic components induced by the electropermeabilization
4.15 pm ──────→ 5.45 pm

The electric field induced permeabilization abolishes the selective barrier that represents the membrane. Since as shown in III.A, exogeneous molecules can be incorporated in the cytoplasm, can endogeneous molecules leak out ? This is shown by measuring the ATP content in the external solution. This molecule is easily detected by use of the luciferin-luciferase assay.

It is thus possible to compare the process of penetration, observed in III.A with those of leaks.

Experimental procedure

1. Non confluent cells are grown on a culture dish.

2. Remove the culture medium with a Pasteur pipet.

3. Add 0.8 ml of the pulsing buffer A.

4. Bring the electrodes in contact with the bottom of the dish by use of the movable stage.

5. Trigger the electric pulses (similar parameters as in III.A).

6. Check the shape of the pulses on the scope during 5.

7. Wait for 1 min. and then suck out 0.5 ml of the solution and put it in an Eppendorf tube.

8. Keep it on ice.

9. Do another experiment with a new culture dish.

ATP assay

A luminescence approach is used. ATP is converted in light by the luciferin-luciferase system. The light emission is directly proportional to the amount of ATP. This emission of light is weak but easily measured with a Scintillation counter.

1. Work with full window opening.

2. Select the "manual, repeated" mode.

3. Select an accumulation time as short as possible.

4. Select the chemioluminescence mode.

5. Use scintillation vials in glass either new or carefully cleaned with sulfochromic mixture and then washed repeatively with distilled water.

6. Add in each vial 0.7 ml of distilled water and 0.2 ml of Glycylglycin buffer 0.2 M pH = 7.4.

7. Cool the vials by letting them standing in the counter.

8. Prepare a solution of Luciferin-Luciferase (20 mg/ml in distilled water), use it fresh, keep it on ice.

9. Add in the vial 0.3 ml of the solution which was in the dish.

10. Add in the vial 10 μl of the luciferin-luciferase mixture.

11. Mix.

12. Lift down the vial in the counting well.

13. Switch on the counter.

14. Run 5 successive accumulations.

15. Extrapolate these counts to the time of mixing (step 11).

16. Convert these extrapolated counts in ATP concentration by reference to a calibration curve obtained under similar conditions.

17. Compare the leak of ATP with the extend of permeabilization you obtained in II.A under the same pulsing conditions.

The ATP assay can be run with a luminometer. The procedures described by the manufacturer should be followed.

D. Cell electrofusion

11.45 am ⟶ 12.45 pm
2.45 pm ⟶ 4.15 pm

Cell fusion can be easily obtained by electric field pulsation. One of the most fascinating properties of "electropermeabilized" cells is their fusogenic character. In other words, if two cells in close contact are electropermeabilized, this induces a mixing of their membranes in the contact area and the appearance of a new "hybrid cell" in which the two nuclei are in the same "cytoplasmic" volume. As the "permeabilization" is fully reversible, the viability of the hybrid cells is not altered.

A high yield of hybridization is observed. In the present experiment, the contact between the cells is obtained spontaneously by the "contact inhibition" which occurs when the cell density on the culture dish is high enough. Pulsing the cell culture under the right pulsing conditions (a function of the cell strain), followed by a 37°C incubation during 1 h. (needed for the cytoplasmic reorganization) lead to a high number of viable polykaryons (cells with a large number of nuclei per cytoplasm).

Experimental procedure

1. Cells are grown on the culture dish up to a density where a large number of cell-cell contacts are present (about 50 to 75 % of the dish surface should be covered with cells).

2. Remove the culture medium with a Pasteur pipet.

3. Add 1.5 ml of the pulsing buffer (A or B)

4. Bring the electrodes in contact with the bottom of the dish by use of the movable stage.

5. Trigger the electric pulsations (5 pulses of 100 µs with field intensity increasing from 0.2 kV/cm is a right choice to begin with).

6. Check, on line with 5, the shape of the pulse on the scope.

7. Lift down the culture dish.

8. Select a new area on the dish which was not previously pulsed (if possible) and run steps 4 to 7 again with a new set of parameters in 5.

9. Remove the pulsing buffer with a Pasteur pipet.

10. Add 1.5 ml of culture medium.

11. Incubate the cells at 37°C during 1 h (at least).

12. Observe the cells under the inverted microscope.

13. The polynucleation index is the percentage of nuclei belonging to polynucleated cells to the total number of nuclei.

$$I = \sum_{2}^{\infty} n\, C_n \Big/ \sum_{1}^{\infty} n\, C_n$$

n number of nuclei per cell
C_n number of cells where n nuclei are present in the cytoplasm

It is deduced from the observation of the cell culture (at least 200 cells must be observed). The results in the pulsed area are compared to the ones in the control areas (a basal polynucleation exists with some cell strains).

14. Plot the extend of polynucleation as a function of the different parameters. Compare it with the extend of permeabilization (part A).
Does the composition of the pulsing buffer play a similar role on the electropermeabilization and on the electrofusion.

Thanks due to Dr Amalric (this institute) for his comments on this manuscript.

V. REFERENCES

Knight D.E. and Scrutton M.C. (1986). Gaining access to the cytosol : the technique and some applications of electro-permeabilization. **Biochem. J., 234** : 497-506.

Neumann E., Schaeffer-Ridder M., Wang Y. and Hoffschneider P.H. (1982). Gene transfer into mouse myeloma cells by electroporation in high electric field. **EMBO J., 1** : 841-847.

Neumann E. (1984). Electric gene transfer into culture cells. **Bioelectrochem. Bioenerg., 13** : 219-223.

Schwister K. and Deuticke B. (1985). Formation and properties of aqueous leaks induced in human erythrocytes by electri-cal breakdown. **Biochim. Biophys. Acta, 816** : 332-348.

Shillito R.D., Saul M.W., Paszkowski J., Muller M. and Potrykus I. (1985). High efficiency direct gene transfer to plants. **Biotechnology, 3** : 1099-1103.

Sowers A.E. (1986). A long lived fusogenic state is induced in erythrocyte ghosts by electric pulses. **J. Cell. Biol., 102** : 1358-1362.

Teissié J. and Blangero C. (1984). Direct experimental evidence of the vectorial character of the interaction between electric pulses and cells in cell electrofusion. **Biochim. Biophys. Acta, 775** : 446-448.

Teissié J. and Rols M.P. (1986). Fusion of mammalian cells in culture is obtained by creating the contact between the cells after their electropermeabilization. **Biochem. Biophys. Res. Comm., 140** : 258-266.

Zerbib D., Amalric F. and Teissié J. (1985). Electric field mediated transformation : isolation and characterization of a TK+ subclone. **Biochem. Biophys. Res. Comm., 129** : 611-618.

Zimmermann U. (1982). Electric field mediated fusion and related electrical phenomena. **Biochim. Biophys. Acta, 694** : 227-277.

Experiment n° 12

USE OF MONOCLONAL ANTIBODY AGAINST KERATIN IN
IMMUNOCYTOCHEMICAL TECHNIQUES

B. DE VEYRAC and P. ADAMI[*]

I. INTRODUCTION AND AIMS

All mammalian cells contain intracellular networks of
protein filaments that can be divided in four types on the
basis of their diameter. In addition to microtubules (22-25 nm
in diameter) microfilaments (5.7 nm) and microtrabecular
filaments (2-3 nm) another group of filaments with diameters
between those of microfilaments and microtubules can be
observed, the intermediate sized filaments (7-11 nm). The
different types can be visualized by the indirect immunofluorescence technique using specific antibodies directed against
is actin, tubulin, or the intermediate filament proteins.

Biochemical and immunochemical investigations have
demonstrated a further subdivision of intermediate filaments
according to their protein subunits, five different types of
intermediate filament proteins can be distinguished, each type
specific for a special group of tissue (Osborn et al., 1981 ;
Ramaeker et al., 1983).

Tissue specificity of intermediate sized filament
proteins

Cell type	Protein constituent
epithelial cells	keratins (cytokeratins prekeratins)
mesenchymal cells	Vimentin
muscle cells	Desmin
neuronal cells	Neurofilament proteins
Astrocytes	Glial fibrillary acidic protein

---

* as immunologist of the workshop

Furthermore recent investigations suggest that the tissue specific intermediate filament proteins are retained during neoplastic transformation (Krepler et al., 1974).

Therefore monoclonal antibodies to these intermediate filament proteins and particulary keratin may be used in neoplastic tissue recognition.

A monoclonal antibody against keratins (KL1) from normal human stratum corneum was obtained using hybridoma techniques (Kohler & Milstein, 1975). KL1 was characterized by its immunohistochemical staining of various epithelia and by its recognition of 55-57 kilodalton keratin polypeptide from normal epidermis using the immuno blot technique (Viac et al., 1983).

Approximatively 80 % of normal keratinocytes isolated after trypsinization were labeled by KL1 whereas most negative cells showed basement membrane zone antigens. This confirmed differences in the expression of medium sized polypeptides between basal and suprabasal cells during the course of human epidermal differentiations. All epithelial cells from other human epidermal tissues (thymus thyroid, bronchial mucosa, stomach, intestines) were positive with KL1 whereas non epithelial cells and tissue did not show any staining. This finding permits differential diagnosis between carcinoma and sarcoma (Caveriviere et al., 1985).

Principle of keratin staining using APAAP technique

Immunotech has developed a kit containing prediluted primary monoclonal antibody against keratin (55-57 KD) and reagents ready to use allowing the alkaline phosphatase anti alkaline phosphatase (APAAP) staining technique (Cordell et al. 1984 ; Mason, 1985). This unlabelled antibody technique is based upon the use of soluble immune complexe formed between calf intestinal alkaline phosphatase and monoclonal antibodies to this enzyme. It is identical in principle to the widely used PAP immuno peroxidase procedure. In the APAAP method the enzyme reaction is developed using a naphthol substrate and Fast Red TR chromogen yielding a red reaction product. Endogenous alkaline phosphatase activity in most human cells is inhibited by the addition of levamisole to the substrate solution.

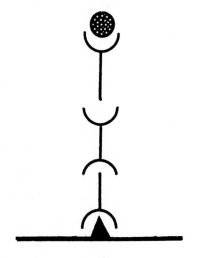

Calf intestinal
alkaline phosphatase   APAAP
                              complexes
Monoclonal mouse
anti alkaline phosphatase

Goat anti mouse Ig

Primary mouse
monoclonal antibody

Antigen

Schematic illustration of the monoclonal APAAP procedure

II. EQUIPMENT, CHEMICALS AND SOLUTIONS

A. Equipment

1. <u>Acess to</u>

Standard light microscope with three objectives 10 x
                                                 20 x
                                                 40 x

2. <u>On the bench</u>

Shaker such as Vortex
Automatic pipettes (5 μl to 1 ml) Gilson
Disposable plastic tubes (5 ml)
Filter paper
Forceps
Plastic gloves
Graduated funnels (10-100-250 ml)
Diamond pencil
Staining jars
Timer
Moist chamber
Absorbent wipes
Slide carrier
Wash bottles

B. Chemicals and solutions

Toluene
Alcohol 100°
Alcohol 70°
Alcohol 50°

Tris HCl (0.05 M pH 7.6) NaCl (0.15 M) (1 liter)
Phosphate Buffer saline pH 7.2 containing 1 g/l of BSA
(1 liter)
Mayer's hematoxylin ⎞
Mounting medium   ⎬   will be supplied by Immunotech
(glycerol gelatin)   ⎠

C. Reagents supplied in Kerato kit 4-0414 (Immunotech)

a) Primary monoclonal antibody against cytokeratin
(55-57 KD) (freeze dried)
        containing :   - monoclonal antibody 0.2 mg
                       - bovine serum albumine 1 mg/ml

b) 10 ml goat affinipure antibody to mouse immunoglobulin prediluted in buffer
     pH 7.6
        containing :   - bovine serum albumine
                     - sodium azide 0.15 M

c) 10 ml Alkaline phosphatase monoclonal anti alkaline
phosphatase immune
     complex prediluted in buffer pH 7.6
        containing :   - bovine serum albumine
                     - sodium azide 0.15 M (APAAP)

d) 10 tablets of alkaline phosphatase substrate
        containing :   - Naphthol MX - phosphate
                     - Fast Red TR
                     - Levamisole

e) 20 ml 0.1 M Tris HCl pH 8.2 Substrate buffer (TBS)

D. Tissue sections

Paraffin embedded sections from normal and malignant tissue.

III. EXPERIMENTAL PROCEDURES

9.00 am 1. Deparaffinization

Removal of paraffin and rehydratation of tissue. It is very important that the embedding medium is completely removed from the specimen, any residual medium will cause an increase in background and thus often obscure specific staining.

* Immerse the slides in two successive baths of toluene for 6 minutes each.
* Transfer the slides in baths of decreasing concentration of alcohol, 100° (5 min), 70° (5 min), 50° (5 min).
* Wash twice in gently running tap water for 30 secondes.
* Label each slide with a diamond pencil.
* Negative controls : it is necessary to evaluate non specific staining and this allows better interpretation of specific staining. One tissue section is intended for negative control.

10.45 am

2. Preparation of reagents
   Monoclonal antibody against cytokeratin

* Restore with 1 ml distilled water. No sodium azide is present.
* Dilute the reagent 1:100 in phosphate buffer saline pH 7.2 Bovine serum albumine 1 mg/ml

Alkaline phosphate substrate solution

The solution substrate is used in step 4 of the staining procedure. The substrate tablets must be dissolved in substrate buffer before use.

* For each tablet to be dissolved, transfer 2 ml from substrate buffer in a tube (one tablet is sufficient for 5 tissue sections).
* Transfer one tablet into buffer using forceps.
* Shake vigourously using a vortex for 1 to 2 min to effect final dissolution.
* Filter the substrate

Solution should be used within 30 min of filtration.

12.00 m 3. Immunostaining procedure

Remove slides from carrier. Carefully wipe away the excess liquid. At no time must the specimen be allowed to dry out. Hydrate deparaffinized tissue section in TBS for 3 min. Remove buffer, wipe slides and quickly proceed to the first step of immunostaining procedure.

Step 1 :   Add 100 to 200 µl of monoclonal antibody to cytokeratin diluted solution.
Incubate in a moist chamber for 30 min at room temperature.
Wash for 1-2 min in TBS (wash bottle) wipe slide.
For negative controls, add 100 to 200 µl of TBS instead monoclonal antibody.

Step 2 :   Add 100 to 200 µl immunoglobulin anti mouse
               solution (yellow bottle)
               Incubate 30 min at room temperature
               Wash for 1-2 min in TBS (wash bottle)
               Wipe slides

Step 3 :   Add 100 to 200 µl of APAAP solution (red bottle)
               Incubate 30 min at room temperature
               It is now convenient to prepare the substrate
               solution for step 4
               Wash for 1 to 2 min in TBS (wash bottle)
               Staining intensity can be enhanced by repeating
               steps 2-3
               Wipe slides

Step 4 :   Add enough drops of filtered substrate solution to
               cover specimen
               Incubate 20 min at room temperature
               Wash in distilled water

Step 5 :   Counter staining procedure -- Cover specimen with
               Mayer's hematoxylin or place slides in a bath of
               Mayer's hematoxylin.
               Mayer's hematoxylin.
               Incubate for 1-5 min.
               Wash the slides under gently running tap water for
               5 min.

Step 6 :   Mounting slides

               Liquify glycerol gelatin by warning at 37°C.
               Cover specimen using liquid glycerol gelatin.

               Slides are now ready for viewing.

17.00 pm   4. <u>Interpretation of results</u>

Slides should be examined by conventional light microscopy. It is necessary to observe at first, the negative control serving as a blank and if any positive staining is detected on this specimen, it is non specific and should be ignored when interpreting results.

For tissue specimen, you can observe staining by the APAAP system resulting in the formation of a bright red precipitate at the site of the target antigen. Hematoxylin stained cell nuclei are in blue.

IV. REFERENCES

Caveriviere P., Al Saati T., Voigt J.J., Delsol G. (1985). Diagnosis of undifferentiated tumors with monoclonal antibodies usable on paraffin sections. **La presse médicale** 28 Septembre 1985, **14**, n° 32.

Cordell J.L., Faleni B.R., Erber W.N., Ghosh A.K., Zainalabidden Abdulaziz, Stuart M.D., Pulford K.A.F., Stein H. and Masson D. (1984). Immunoenzymatic Labeling of Monoclonal Antibodies using Immune complexes of Alkaline Phosphatase and Monoclonal Anti alkaline Phosphatase (APAAP complexes). **J. Histochem. Cytochem. 32** : 219-229.

Kohler G., Milstein C. (1975). Continuous cultures of fused cell secreting antibody of predefined specificity. **Nature,** (London) **256** : 495-497.

Krepler R., Denk H., Weirich E., Schmid E. and Franke W.W. (1974). Keratin like proteins in normal and neoplastic cells of human and rat mammary gland as revealed by immunofluorescence microscopy. **Differenciation 20** : 242-252.

Mason D.Y. (1985). Immunocytochemical Labeling of monoclonal antibodies by the APAAP immuno alkaline phosphatase technique. In **Techniques in Immunocytochemistry** Vol 3.

Osborn M., Geisler N., Shaw G., Sharp G. and Weber K. (1981). Intermediate filaments. **Cold Spring Harb, Symp. quant. Biol. 46** : 2490-2494.

Ramaekers F.C.S., Puts J.J.G., Moesker O., Kant A., Huymans A., Haag D., Jap P.H.K., Herman C.J. and Vooijs G.P. (1983). Antibodies to intermediate filament proteins in the immunohistochemical identification of human tumors : an overview. **Histochem. J. 15** : 691-713.

Viac J., Reano A., Brochier J., Staquet M.J. and Thivolet J. (1983). Reactivity Pattern of a monoclonal Antikeratin antibody. **J. Invest. Derm. 81** : 351-354.